Praise for *Citizen Outlaw*

"A fine book about a New Haven gangleader tu~~rned~~ ~~co~~ach officer who, by talking down revenge-m~~inded~~ ~~figu~~res to interrupt the cycle of retaliati~~on~~ ...

—*~~New Yorker~~*

"A quintessentially New Haven ~~story~~ ... ~~a d~~enunciation of the violence and errors of his yout~~h~~ ~~and h~~is inspirational striving toward an almost improbable redemption."

—*New Haven Register*

"A great book, a great writer."

—Paul Bass, *New Haven Independent*

"A remarkable story."

—Colin McEnroe, WNPR

"Inspirational."

—*Kirkus Reviews*

"An elegantly written account tackling one of the most important moral issues of our time—the role of redemption in our justice system and our society. This book offers an unforgettable rendering of a life redeemed."

—Jennifer Gonnerman, author of *Life on the Outside: The Prison Odyssey of Elaine Bartlett*

"Inspiring. . . . Barber could not have chosen a better subject; Outlaw is smart, charismatic, and someone readers will instantly root for. . . . A must-read especially for those interested in social justice and prison reform."

—*Booklist*

"Plunges readers inside the belly of the beast with such searing credibility you'll feel you're there—witnessing a gangland murder, the brutal inhumanity of prison life stripped of Hollywood fluff, and, ultimately, a man's redemption and call to help others lost on our cities' mean streets. Brilliantly crafted, *Citizen Outlaw* is that rare book that educates, entertains, and inspires. It's a must-read for anyone interested in criminal justice reform and the power of the human spirit."

—Pete Earley, author of *The Hot House: Life Inside Leavenworth Prison*

"An extraordinary book about an extraordinary man. *Citizen Outlaw* is a great American gangster story, told with wit and insight. William Juneboy Outlaw III would have undoubtedly made his mark no matter where he was raised, but his home is New Haven, a city to which he has been, at various times, a curse and a blessing. On the page Juneboy is excellent company, and this account of his life is riveting and unexpectedly moving—a life-and-death saga that is also a meditation on the nature of ambition."

—Kelefa Sanneh, staff writer, *The New Yorker*

"What does it mean when the person least likely to turn around his life actually does it? William Outlaw's story, told by Charles Barber with sensitivity and wisdom and jaw-dropping detail, will amaze and inspire you."

—Robert Kolker, author of *Lost Girls: An Unsolved American Mystery*

CITIZEN
OUTLAW

CITIZEN OUTLAW

ONE MAN'S JOURNEY FROM GANGLEADER TO PEACEKEEPER

CHARLES BARBER

ecco

An Imprint of HarperCollinsPublishers

CITIZEN OUTLAW. Copyright © 2019 by Charles Barber. All rights reserved. Printed in the United States of America. No part of this book may be used or reproduced in any manner whatsoever without written permission except in the case of brief quotations embodied in critical articles and reviews. For information, address HarperCollins Publishers, 195 Broadway, New York, NY 10007.

HarperCollins books may be purchased for educational, business, or sales promotional use. For information, please email the Special Markets Department at SPsales@harpercollins.com.

Ecco® and HarperCollins® are trademarks of HarperCollins Publishers.

A hardcover edition of this book was published in 2019 by Ecco, an imprint of HarperCollins Publishers.

FIRST ECCO PAPERBACK EDITION PUBLISHED 2020

Designed by Joy O'Meara
Photographs are courtesy of the author unless otherwise credited.

Library of Congress Cataloging-in-Publication Data

Names: Barber, Charles, 1962- author.
Title: Citizen Outlaw : a gangster's journey / Charles Barber.
Description: First edition. | [New York, New York] : Ecco, [2019] | Includes bibliographical references.
Identifiers: LCCN 2019013862 (print) | LCCN 2019980344 (ebook) | ISBN 9780062692849 (hardcover) | ISBN 9780062692856 (paperback) | ISBN 9780062692870 (ebook)
Subjects: LCSH: Outlaw, William Juneboy, III. | African American criminals--United States--Biography. | Community activists--Connecticut--New Haven--Biography. | Crime prevention--Connecticut--New Haven--Citizen participation--Biography. | African Americans--Social conditions. | United States--Race relations.
Classification: LCC HV6248.O B37 2019 (print) | LCC HV6248.O (ebook) | DDC 364.1092 [B]--dc23
LC record available at https://lccn.loc.gov/2019013862
LC ebook record available at https://lccn.loc.gov/2019980344

20 21 22 23 24 LSC 10 9 8 7 6 5 4 3 2 1

For Laura Radin and John Barber

and

in memory of Thomas Ullman, the Atticus Finch of New Haven

—Charles Barber

For my children and Germaine, my everything

—William Outlaw

Every saint has a past and every sinner has a future.

—OSCAR WILDE

AUTHOR'S NOTE

The narrative that follows is William Outlaw's individual story, which brought him from New Haven, to prisons around the country, and finally a return to his hometown. This account is not a definitive history of the Jungle Boys gang nor a social history of New Haven.

For purposes of privacy, many of the names of individuals and certain identifying details have been altered.

CONTENTS

Part Two CITIZEN

SEPTEMBER 24, 1988

William Outlaw's first mug-shot, 1984.
New Haven Police Department.

WILLIAM OUTLAW SAW, THROUGH THE FAINT GLOW OF THE STREET-lights of Church Street South, a man approaching him. He would learn later that the name of this man was Sterling Williams. The man came closer to Outlaw, and Outlaw could see that this man, Sterling Williams, was stocky. To Outlaw, who was twenty at the time, the encroaching man appeared to be in his thirties.

"I'm here for Juneboy," Sterling Williams said. He spoke with a Jamaican accent. Juneboy, William's middle name, was also his street name, and what everybody in New Haven called him. His full given name was William Juneboy Outlaw III.

Outlaw said nothing to the man—he just examined him further—but he felt for the revolver in the deep pocket of his cargo pants.

As Outlaw surveyed the interloper before him, it was one in the morning and unseasonably warm in New Haven, Indian summer. Outlaw was sitting on the hood of an abandoned car, an old Ford, holding court, as usual. He was surrounded by five young men, teenagers mainly, and all members of his gang, which was called the Jungle Boys. The young men circled around Outlaw wore Nike and Adidas sneakers, camouflage pants and green T-shirts, for green was the official color of the Jungle Boys, and the more prosperous among them wore gold chains. They stood on the edge of a run-down and densely packed housing project known as the Jungle, which Outlaw had selected three years earlier as the base of his criminal operations, the perfect place from which to deal cocaine. As Outlaw spoke, his gold-plated front teeth flashed under the streetlights. One tooth was engraved with the letter *J*, and the other with the letter *B*.

As he sat on the hood of the abandoned car, the various members of the Jungle Boys listened to him attentively. Outlaw was their indisputable leader, renowned equally for his loyalty to them and his viciousness to rivals. For most of the members of the gang, which numbered about forty, Outlaw had represented their meal ticket out of poverty, which was pervasive through large swaths of the city, especially among young black men. Virtually all of New Haven's factories had long since closed, and being selected by Outlaw for the gang was for some of them like winning an unexpected lottery. There was something irrepressible about Outlaw, a quality of complete self-possession: when he talked to you, it was as if no one else in the world even existed.

As Sterling Williams stood there, and Outlaw gripped his gun wondering what he would do next, the night air was almost misty. The occasional car drove by on Church Street South and there was the faint rumble of traffic from Interstate 95, a quarter mile away.

Outlaw and the Jungle Boys had been expecting a fight all day. Outlaw had heard earlier that day that Sterling Williams or someone like him might be visiting—there had been rumors flying around the Jungle that there might be trouble coming from New York, in the form of Jamaican gangsters looking to shoot, kill, and torture, whatever was necessary, to force themselves into new territory. Within the last year, the Jamaicans had taken over most of the drug trade in Bridgeport, a city twenty miles to the west. A few months earlier, in an opening salvo in their bid to dominate New Haven, they had killed a childhood friend of Outlaw's, a lone-wolf drug dealer. The Jamaicans had shot him execution-style, five bullets to the head, up at East Rock Park, at the top of a cliff that gloomily overlooks the entire city. The killing was interpreted by gangs in New Haven as largely symbolic, intended to send a message to the city that the Jamaicans were ready to take over.

Earlier that day, one of the Jungle Boys had been playing rap music, loudly, on a boom box, while proceeding down Liberty Street a few blocks from the Jungle. A young Jamaican with dreadlocks had accosted the member of the Jungle Boys and told him to turn off the music. The Jungle Boy, trained to be on the offensive, had taken a pistol out of his pocket and smacked the Jamaican with the butt of the gun, repeatedly, until he crumpled to the pavement. Blood already beginning to drip from his face, the Jamaican had uttered, "We're going to fuck all of you guys up. We're going to kill Juneboy."

This encounter prompted a chain reaction of events, conducted with the high efficiency with which the Jungle Boys, now three years in operation, conducted their business. Outlaw was paged by his second in command and protégé, Ricky, who helped run the gang while Outlaw was out of town. Outlaw was in Harlem that day, ostensibly on business, but also pleasure. He was procuring cocaine, the lifeblood of the Jungle Boys, kilos of it from the Albanian intermediaries with whom Outlaw had been buying for years. But he'd also been partying there, on 125th Street in particular, taking advantage of the clubs, and women, and clothing boutiques that he had begun to frequent as the Jungle Boys had grown in power and resources. Ricky told Outlaw to come back to New Haven immediately. They needed his help. Driving one of the forty cars that the Jungle Boys now maintained back home, he had pulled up to the Jungle fifteen minutes before Sterling Williams appeared.

Now, Sterling Williams stepped even closer toward the group of Outlaw's boys and repeated, "I am looking for Juneboy."

Outlaw was on alert, but he wasn't particularly anxious. He wasn't generally an anxious person. But he gripped the handle of the gun in his pocket harder.

One of his underlings, a slim eighteen-year-old looking to work his way up in the gang, said, "I'm Juneboy."

Outlaw liked that—it gave him cover. That's what he had trained his guys to do. And it allowed Outlaw to get closer to Sterling Williams without Williams being aware. Outlaw moved within five feet of Sterling Williams, so that he could see his dark face through the night shadows. Williams was wearing a baseball cap backward and had long dreadlocks.

Sterling Williams said, in a lilting Jamaican accent, and to the wrong guy, "I'm going to fuck you up. I am going to take over your gang. I'm from New York and I'm going to kill all of you, and I'm going to kill Juneboy."

Outlaw was shocked at the man's sheer and utter recklessness. He could not believe his aggressiveness, nor his foolhardiness. No one ever addressed him, or a person presumed to be him, like that, ever. The Jungle Boys did the terrorizing, not the other way around. They were by far the biggest gang in New Haven, and the only one to operate in three neighborhoods. Outlaw's crews were organized by shifts, job descriptions, and pay scales. Last year he'd made something like a million dollars, although he wasn't really counting.

Outlaw figured that, in order to be so crazily reckless, this man Sterling Williams must have some serious backup. Outlaw didn't know much about the Jamaican gangsters coming out of New York, but he did know that they called themselves the Shower Posse, and that they were especially vicious, even by a gang's standards. They performed ritual killings, tortured people, killed your family if they couldn't get you. He knew that they were based in Bedford-Stuyvesant and he knew that two splinter groups of the Shower Posse, calling themselves the Rats and the Cats, had taken over Bridgeport. Outlaw peered through the darkness to the small apartment buildings and ramshackle ranch houses and parked cars on the other side of the street. He expected to see more guys, waiting to attack. But all was quiet. There didn't seem to be anybody else.

"I'm gonna fuck you guys up," Sterling Williams said again to the man he presumed to be Outlaw.

Outlaw found himself starting to seethe.

A second man appeared through the shadows. Later Outlaw learned that this man's name was Fitzroy Phillip. He was also dark-skinned with dreadlocks, and a little older and stockier, probably in his mid-thirties. Maybe this was the beginning of the onslaught, Outlaw thought. Maybe more really were coming now. Maybe they were going to come in groups, waves.

Sterling Williams was now just a few feet away from Outlaw.

"I'm going to destroy you guys! Take you the fuck ov—"

Williams never finished the sentence because in that moment his skull exploded. He fell to the ground instantly. Outlaw had shot him in the head. Williams never saw it coming. He fell immediately and a massive pool of blood instantly oozed over the sidewalk.

In the chaos, Fitzroy Phillip sprinted away. He ran toward the modest, run-down houses across the street. As Phillip was halfway across Church Street, Outlaw fired his revolver, clipping Phillip in the leg twice and three times in the back. Phillip staggered but somehow managed to keep running until he was out of sight in the darkness.

The Jungle Boys looked at one another.

"Let's get the fuck out of here," Outlaw said to his guys, and they all scattered in different directions—except for Ricky who stayed with Outlaw. That's what Ricky did—he stayed with Outlaw at all times, never leaving his side. Ricky and Outlaw ran through the housing project to a car, a 1977 white Buick Riviera, that they kept parked at the back of the Jungle for just such occasions. Ricky found the keys that were kept under the seat. Outlaw jumped into the passenger seat and Ricky drove the car off into the night. As they drove, Outlaw felt his gun in his hand. The barrel was still hot to the touch.

Sometime later that night, when the police collected Sterling Williams's body off the pavement, they found two guns and approximately one thousand dollars cash on his person.

———

FORTY-FIVE MINUTES AFTER the shootings, Outlaw and Ricky sat at Howard Johnson's on a service road parallel to Interstate 95, which runs through New Haven on its journey from Maine to Florida. The New Haven harbor and Long Island Sound were a quarter mile from where they sat, in a booth in the bright, pleasant restaurant, listening to Muzak, eating cheeseburgers, and talking to the waitress.

Outlaw wasn't particularly concerned about what had just transpired in the Jungle. He felt no particular remorse or anxiety. There were so many shootings in New Haven—twenty-five homicides in the last year, a remarkable number for a city of 130,000, double the homicide rate of Boston or New York—that it was hard for the police to even keep up. In Outlaw's experience, the police were less interested in or inclined to investigate murders that happened in the projects. Besides, it was self-defense—the Jamaicans had invaded his territory and threatened him. He had performed a service, actually, preventing the posses from moving into New Haven like they had Bridgeport. No, there was nothing in particular to worry about.

Ricky followed Outlaw's lead as they made small talk. Ricky looked up to Outlaw. Outlaw, at twenty was huge: six foot four, three hundred pounds, most of it pure muscle. Ricky was a favorite of Outlaw's. Ricky was like a slightly older brother, and he was bright and quick. Recognizing Ricky's energy and natural aggressiveness, Outlaw had groomed him for a couple of years now. Ricky had dropped out of high school for the opportunity.

Howard Johnson's was a police hangout, which was one of the things that Outlaw liked about it. He preferred to keep tabs on the police; it was one of his specialties as a gangster. In fact, the Jungle was only two hundred yards from the New Haven police station, and a lot of people thought Outlaw had been crazy for setting up his gang under their noses. But Outlaw figured the more he knew about the police, the better. If they were going to investigate him, he could investigate them too. Sometimes he would go out to the police parking lot at midnight, at shift change, and stand there ominously and watch as the officers got into their cars. Sometimes he would even follow them home to the leafy suburbs outside

New Haven. He wanted them to know that he knew where their wives and children lived. That made him feel good. The cops tended to back off after those little excursions.

Now, two police officers walked into the restaurant and took a booth fifteen feet away from where Ricky and Outlaw sat. The officers ordered coffee. Ricky and Outlaw became quiet. Outlaw listened to their scanner and heard, distinctly: "A homicide at Church Street South. William Juneboy Outlaw the chief suspect."

Outlaw looked at Ricky. Ricky nodded. Outlaw peeled off two twenty-dollar bills from the rolls of cash, tied in rubber bands, which he kept in his pocket to pay for the meal. He and his inner circle typically carried rolls of $5,000 or $10,000. They got up quietly and left the restaurant. The policemen didn't seem to notice them.

Outlaw and Ricky walked out to the parking lot. In front of them was the harbor, with a gloomy view of the water, and behind them the warehouse district. The meatpackers were arriving for the early shift. Ricky and Outlaw didn't say much. Ricky said, "Best of luck, Juneboy," and watched as Outlaw got into the white Buick, drove to a nearby entrance ramp, and then merged onto Interstate 95, heading to New York City.

A few hours later, having driven the two hours from New Haven to Manhattan, William Outlaw threw the gun that killed Sterling Williams into the murk of the Hudson River somewhere near 125th Street in Harlem. Half an hour after that, he checked himself into the Novotel, a four-star hotel in Times Square, for what would turn out to be a weeklong stay.

Twenty-four hours later, two New Haven police officers, Joseph Pettola and Gilbert Burton, approached a man in his bed at Yale New Haven Hospital. Fitzroy Phillip was connected to various machines, wounded from the five bullets that had hit him in the leg and the back, but awake and alert.

Pettola and Burton had prepared a ring binder containing photographs, mug shots, mainly, of 250 young black males. They placed the binder on a tray on Phillip's chest and flipped the pages for him, as he was unable to do so himself. They asked Phillip to identify the man who had shot him and killed Sterling Williams. Phillip said nothing during the process until Burton and Pettola came to one particular photograph. Phillip pointed to it and said, without hesitation, "This is the guy."

"Are you sure?" said the detectives.

"One hundred percent," Phillip said, in his Jamaican accent.

LOOK AT IT on the map, and the city of New Haven sits at the top of a decisive triangular gouge set three miles back from the prevailing coastline of Long Island Sound. It is an almost perfect natural harbor. The Native Americans called the region Quinnipiac, meaning "long water place." It is not surprising that the first white settlers, a group of Puritans fleeing the Massachusetts Bay Colony, sought the harbor as a place of refuge in the winter of 1638. After surviving that first brutal winter, they gave the city its name in gratitude. Today the Port of New Haven, on the east side of the harbor, is the busiest port between New York and Boston. There are steep moorings for tankers and other shipping vessels, and hundreds of warehouses and storage tanks containing tens of millions of barrels of petroleum products. In the warehouses are stored general bulk items, cargo, scrap metal, metallic products, cement, sand, stone, and salt.

But for the many thousands of people who drive daily on Interstate 95 through New Haven, it is hard to discern any of this. The highway effectively cuts the city off from the port and the ocean, leaving only fleeting glimpses of Long Island Sound. Instead, commuters encounter a haphazard maze of flying exit ramps, aging corporate office parks, gas stations, isolated parking lots, and a defunct sports betting center. There is a good fish restaurant but it is hard to get to. There is even a small stretch of forgotten beach, but no one would ever think of swimming there. New Haven is famed for its pizza, thought by some to be the best in the country, and more consequentially for Yale University and its uber-elite reputation, soaring Gothic towers, gated quadrangles, and $27 billion endowment. Yale, however, is but a small part of the city, geographically and spiritually. Gritty, blue-collar industry has been the hallmark of New Haven for most of its existence. The city is where Eli Whitney developed the cotton gin, where Samuel Colt invented the revolver, and where Oliver Winchester refined the design of the repeating rifle—"the gun that won the West," as well as two world wars.

It is surprising, the poverty that exists in New Haven. Connecticut, by a number of measures, is the wealthiest state in the country. New Haven, along with Hartford, counts as the state's most prominent city, and Yale is the second-richest university in the world. New Haven has one of the largest hospitals on the East Coast, Yale New Haven; a thriving biotech industry; and excellent restaurants and theater. The city is surrounded to the west and north by wealthy and woodsy suburbs populated by engineers, doctors, lawyers, and professors, and to the east by picturesque seaside towns with bookstores, antiques shops, and art galleries. But about 30 percent of the city's residents live below the poverty line, and three of New Haven's neighborhoods—Newhallville, the Hill, and Dixwell—are among the poorest in the country. In the Hill more than 40 percent of families of four live on less than $24,000 a year.

What makes the depth of the privation all the more shocking is that New Haven, which reached its peak population in 1950, is no longer that big a city, but it has problems and decay that one would associate with a much larger urban center. The shrinking population, now around 130,000—thirty thousand less than in 1950—is evidenced by the growing number of conspicuous abandoned houses with smashed-out windows and sagging porches. In 2011, New Haven was ranked as the fourth most violent city in the country by *The Atlantic*. "New Haven has historically had the highest rate of violent crime on the East Coast," the citation noted. "[The city] has the eighth-highest rate of robbery and the fourth-highest rate of assault in the U.S." The vast majority of the victims, as well as the perpetrators, of the homicides are young black men. In 2011, every single victim of homicide—and there were twenty-four of them—was a person of color. The racial dynamics of the city have turned upside down in the last fifty years: in 1970, whites comprised 70 percent of New Haven's residents; now the figure is 42 percent.

A melancholy "betwixt and between" feeling pervades the city. New Haven is about midway between Boston and New York; it is a port, but you can't see the water; it is the site of one of the world's most famous universities, but most of the city's residents have nothing to do with it. New Haven used to be called "the Elm City" but now all the elms are gone, killed off by disease. Even on the sunniest days, New Haven's perfect natural harbor is a brooding steel gray. The detritus of the city's former industrial activity seems to have slid into the harbor, permanently fouling the waters, even if all the factories are long gone, simply abandoned or converted into condominiums with large For Sale signs visible to commuters as they drive rapidly past on the highway.

William Juneboy Outlaw III grew up in Quinnipiac Terrace, a housing project three miles north of New Haven harbor. Quinnipiac Terrace, or Q View, as its residents called it, was a hundred yards from the Quinnipiac, a river that flows through the city to the Long Island Sound. Q View opened in 1941, and was comprised of a series of rectangular barracks-like buildings that rose starkly out of the earth with no landscaping around them. Outlaw lived in one of the complex's 248 units with his hardworking and long-suffering mother, Pearl, and his older brother and sister. Outlaw had little in common with his siblings and had little to say to them. Their father had left the household before Outlaw was born. Q View was a place of supreme isolation, almost quarantined from the adjoining neighborhood of Fair Haven. The project felt like an out-of-the-way afterthought to house the city's poor. The residents of Q View were virtually all African American, and whenever Outlaw ventured into Fair Haven, which was more mixed, motorists would scream at him, "You're black as tar!" or "Go back to Africa!" On his way to school one day, he stopped into the fire station and asked for a drink of water. "Fuck you, nigger," responded the fireman.

Even as an elementary school student, Outlaw was vastly bigger than his peers, often a head taller and thirty pounds heavier. A schoolyard basketball and football star, he was the first to be picked as lineups were formed. His size was so conspicuous that he was constantly goaded into schoolyard fistfights. "Juneboy! Juneboy! Juneboy!" groups of boys and girls would chant, seeking live entertainment. Outlaw was only too happy to oblige. Usually the skirmishes would not last long: a few introductory punches until Outlaw landed a hard shot and his opponent would collapse. It was so entertaining for others, and so generally easy for Outlaw, that he often fought three or four times a week. But school itself was an optional consideration. Outlaw did not know a single person who had gone to college, and few who had graduated from high school. Many mornings, Outlaw and his friends would head to school, only to veer off to do something more interesting, like play ball or steal potato chips from delivery trucks when the driver wasn't looking.

When Outlaw was ten, some older boys, sitting around a picnic table at Q View, introduced him to pot. After some initial tutoring, Outlaw with his usual zest took drag after drag on the joint until he became dazed and dizzy. The boys laughed, thinking it hilarious how bloodshot his eyes were. Initiation in another form came just a year later. Older girls, who just a few years earlier had played hopscotch or King of the Hill with Outlaw, forced themselves upon him in the shadowy corners of Q View. Over time, Outlaw was an increasingly willing participant in these activities, and by the time he was twelve, he'd had sex with five different girls.

At night in his mother's apartment, Outlaw would try to read history books—he had always loved figuring out how things got to be the way they were—or listen to the Boston Celtics on the radio. To make ends meet, Pearl worked two full-time jobs: by day on the production line at the Winchester gun factory, and at night at a linen manufacturer. In the small bedroom that he shared with his brother, and with whom he seemed to share almost nothing else, Outlaw would sometimes hold his head under his pillow to block out the sound of drunken fights and cursing in the unit's hallways. Leaving for school in the morning, he stepped over used syringes. In theory there were housing police at Q View, a special branch of the New Haven police charged with maintaining order over the premises. At age seven, Outlaw so much admired one of the officers, a strapping man who would go on to become New Haven's first African-American police commissioner, that he wanted to become a policeman. But by age ten, Outlaw had seen the behavior of other officers, who took bribes and slept with the single moms, and he grew to hate everything about the New Haven police department.

Q View was at the exact point where the river takes a lyrical turn, where the Quinnipiac for a brief stretch widens dramatically to perhaps a hundred yards across with an additional fifty feet of marshes with tall, swaying grass on both shores of the river. As a boy, Outlaw spent as much time as he could looking out upon the river. Unable to afford a fishing rod, he fished its waters for bunker with a stick and a string, which he sold to the older men for bait. It was his first entrepreneurial activity.

The water was always his salvation. If he had a bad day at school or if his mother was upset with him or if his absentee father showed up to beat him with a belt—all of which happened often—Outlaw would go to a spot under the highway where a channel of the river flowed two hundred yards away from the apartment. Outlaw would lie down on a concrete platform next to the water, and look at the undergirding of the highway twenty feet above his head, and hear the whoosh of the traffic, and feel the vibrations of the cars and the trucks. He would look out over the water and dream about being somewhere else. Perhaps North Carolina, where his grandmother lived, in a wooden shack on a dirt road in tobacco country. Or the Caribbean. It didn't matter where it was, just as long as it was different from where he was now. He would think about a place that was quiet and distant. He would think of another world, pure and clean and watery, and, most critically, far away.

A few years after that, Outlaw was beset by other kinds of visions. At age thirteen, as the result of his role in an armed robbery, he was sent to a juvenile detention center where he was surrounded by young boys his age. These were kids from the so-called gritty cities of Connecticut: New Haven, Bridgeport, Hartford, and Waterbury were poor and dangerous areas in a state that was otherwise wealthy, and in some areas, preposterously so. Most of the kids were in for petty crimes—simple robberies and carjackings, mainly. All of them, it seemed, wanted to go right back to that life as soon as they were released.

But at the detention center, Outlaw didn't feel like he was similar to the other boys. He felt that he was cut from a different cloth. The main thing he took from his time at the facility was that he seemed to want things—lusted for them and was willing to work hard for them—more than most people did. He wanted big things, things to match his outsize physique and ebullient personality. Money. Fame. Power. No one in his family had ever been wealthy or famous. Outlaw was going to undo all that, reverse it with sheer willpower. He was going to have Cadillacs and gorgeous women and custom-made suits and swashbuckling hats, and he was going to run half of New Haven. It was going to be uproarious, dream-like, and insane. Soon enough, he would walk into the Apollo Theater in Harlem, or drive around in tricked-out Porsches and Mercedeses, and everyone would know who he was. In just a few years, people were going to look at him, the way he was dressed, the gold chains he was wearing, the small army of young men around him who feared him, and they were going to say, "Who the hell is that guy?"

THIS BOOK TRACKS the improbable, impossible lifetime journey of William Juneboy Outlaw from founder and leader of one of the most brutal gangs in New Haven's history to a powerful force for good in the same community. It is a journey that has taken him all over the United States and it is a journey that by all rights he should not have survived, both psychologically and physically. Outlaw still carries a bullet in his skull, but the devastation to his psyche is much greater. At the heart of his story is an unrelenting paradox. It is a story of doing unremitting damage and then trying to undo that damage, which of course is not actually possible. But these days Outlaw relentlessly spends his hours undertaking the insurmountable task of trying to overturn what has already been done.

Outlaw's journey also embodies a distinctly American narrative, the capacity for self-reinvention. He is the ultimate rebuttal to F. Scott Fitzgerald's maxim that "there are no second acts in American lives." Outlaw has refuted multiple times over what Fitzgerald wrote. For this is the story of a man who was able to successfully negotiate an exit plan from hell.

CITIZEN
OUTLAW

Part One

OUTLAW

With all my niggaz, all my guns, all my women
Next two years, I should see about a billion.

—"The Notorious B.I.G." Biggie Smalls

CHAPTER 1

The Money, the Fame, and the Power

William Outlaw, age eight, New Haven.

THE FIRST ROBBERY WASN'T EVEN HIS IDEA.

In 1981, when William Outlaw was thirteen, Pearl moved the family from Q View to a recently constructed townhouse on Shelton Avenue in Newhallville, in the heart of New Haven. The new townhouse had three bedrooms and new appliances and gleaming wood floors. For Pearl, it was a triumphant move, delivering the family from the projects. For Outlaw, it was like moving to Manhattan from the hinterlands of New Jersey. The central blocks along Newhallville's main artery, Dixwell Avenue, were populated with pharmacies, gas stations, diners, and hairstylists, and adjoined by leafy residential side streets.

But the main reason for Pearl's move was to be closer to the world-famous Winchester gun factory, where she was now a forewoman. The factory, a few blocks from the new townhouse, was a small city unto itself. During World War II, 21,000 people had worked there, and to meet demand, Winchester's managers, against the objections of white employees, advertised in the South, inducing generations of African Americans to come north. By the 1980s, Winchester was beginning the end of its run—the company would cease production in 2006—but for a single mom like Pearl without a high school education, the job paid well and had excellent benefits. A few years later, when Outlaw was shot multiple times during his gang career, he was sometimes treated at Yale New Haven Hospital on his mother's superb Blue Cross Blue Shield insurance.

But for all the advantages of the move, Newhallville was still among the poorest and most violent parts of the city. Newhallville's houses, harkening back to its prosperous industrial past, were large and sprawling, but now few single families could afford an entire home, and the dwellings were divided up into as many as six apartments. A good number of the houses hadn't been painted for years, and the small yards were often untended and overgrown. Certain blocks, such as Lilac Street, not far from Pearl's townhouse, were famous as drug marts. Newhallville was less than three-quarters of a mile from Yale, but the university may as well have been a continent away. Virtually no one in Newhallville—unless they worked a maintenance or kitchen job at the university—had any reason to go to the campus, and the vast majority of Yalies, had you queried them, would have no idea that Newhallville even existed.

Ever practical, Pearl moved the family over the summer so as not to disrupt the school year for her children. Outlaw would start eighth grade in the fall. From the moment of his arrival in Newhallville, Outlaw caused commotion. At first no one particularly knew him: in those days, before the advent of social media, New Haven was still a city of strict geographical neighborhoods, largely isolated from one another. But Outlaw, now approaching six feet and well over two hundred pounds and developing a penchant for flamboyant neon-colored clothes, was an instant mark. He was constantly challenged to fistfights, and dispatched opponents with the same efficiency as he had at Q View.

One of the first of the boys that Outlaw met was Stevie Saunders, a year older than Outlaw, which was just enough of a difference that Outlaw looked up to him. Stevie was a mouthy, energetic boy of medium height, and built like a wrestler. He was current on the latest rap and soul music; wore a thick gold chain around his neck, the only kid in the neighborhood to do so; and kept himself in great shape. Outlaw took to hanging out with Stevie on Dixwell Avenue, flirting with girls, and mixing it up with older boys and young men. Outlaw saw that Stevie had access to a steady stream of expensive possessions—designer sweat suits, the latest Nike and Adidas sneakers, a series of new gold chains—which was surprising, because his family had little money. In bits and pieces of conversation, Stevie revealed to Outlaw that he was working as a stickup artist, robbing stores and individuals around town, armed with a gun.

One afternoon, Stevie suggested to Outlaw that they go in on a job together. Up until that point, Stevie had undertaken his ventures as a solo operator, but no doubt he was impressed by Outlaw's fearlessness. "Don't worry," Stevie assured Outlaw, "I'll lead this." Stevie laid out the plan with an appealing precision. They would go in on the scheme on the next Friday, which was payday. Summer workers, Stevie explained, often cashed their checks at the back counter at Visels Pharmacy, a popular establishment on Dixwell. The plan was for Stevie and Outlaw to wait outside the pharmacy, look for a good victim (someone suitably mild mannered who appeared unlikely to retaliate), follow them after they left the store, jump them at gunpoint, and get the cash. Outlaw thought it was a great idea. The notion of using a gun, the possibility of confrontation: neither gave him pause. Stevie was so convincing that Outlaw thought nothing could go wrong.

On the appointed afternoon, Stevie met Outlaw at five near Visels. Stevie took a gun out of his gym bag and handed it to Outlaw. It was a sawed-off shotgun, a terrifying-looking thing. Stevie explained that

Outlaw would be the gunman, which Stevie had not specified earlier. Outlaw had never shot a gun before, but he placed the firearm under his heavy jacket. The boys stood near Visels and watched as customers went in and out. Eventually Stevie pointed to a man in his thirties, wearing paint-stained coveralls and cashing his check inside. He was of a reasonable stature, but something about him seemed soft. He projected some kind of ineffable vulnerability.

Stevie and Outlaw watched as the man placed the cash in his wallet and left the pharmacy. Keeping a thirty-foot distance, they followed him as he entered a quiet side street. Perfect. Halfway down the block, as the man approached a large brick building with a parking lot adjacent to it—what everybody called the "state building," as it was the home of the Department of Social Services—Stevie sprinted ahead and turned around to face the man. "Walk over there, motherfucker," Stevie said, pointing to the parking lot as Outlaw pulled up by the man's side, flashing the shotgun.

"Hand over the money," said Stevie, in B-movie fashion. The man did as he was told, his hands quivering. Outlaw felt almost relaxed during the process. It was all kind of fun. Stevie shouted at the man, "This never happened!" and then he and Outlaw sprinted away. They turned into a shadowy driveway, where Stevie counted the money. Four hundred dollars. He handed $200 in crisp bills to Outlaw, who in turn gave the shotgun back to Stevie. They laughed and celebrated together, trying to catch their breath.

That night, Outlaw bought weed from one of the many dealers on Lilac Street. Lying in bed as midnight approached, pleasantly stoned, he started laughing once again. He couldn't get over the look on the man's face when he'd seen that crazy-looking gun. His bugged-out eyes were hilarious. But the thing that most cracked Outlaw up was that the gun hadn't even been loaded. Stevie had told Outlaw that after the robbery.

But then Outlaw heard his mother coming home from the night shift at the gun factory, her feet shuffling around in the kitchen, making herself some tea. Outlaw stopped laughing and stashed the cash under his mattress. As he drifted off to sleep, he smiled with content-ment. He loved the feeling of power that came with that gun, loaded or unloaded.

ALTHOUGH THIRTEEN-YEAR-OLD WILLIAM Outlaw could not necessar-ily conceive of it in this way, at the very moment that he took the gun from Stevie, he joined a family tradition of violence that had flowed like a powerful river through the Outlaw family from one generation to another.

For in the exact moment that Stevie and Outlaw stood outside Visels, William's grandfather—whom Outlaw had never met—was in a prison cell in northern Connecticut, serving a twenty-five-year sentence for murder. The murder sentence was merely the finishing touch on a lifetime of sporadic violence. The reason that his grandfather, William Juneboy Outlaw I, had come to Connecticut from North Carolina in the first place was to escape the police, who were seeking him on domestic vio-lence charges after he had beaten his wife. Outlaw's grandfather became a factory man in New Haven, and on payday he would go to a bar with a friend and drink himself senseless. He would wake up the next day and find himself robbed of all his cash. It took Outlaw's grandfather some time to realize that his drinking friend was the perpetrator, but when he did, he abstained from alcohol one Friday night and shot his friend dead in the heart.

Outlaw's father, William Juneboy Outlaw II, worked for the New Haven railroad, repairing track. He was good-looking and tall, a ladies' man. On his days off he wore colorful suits and drove a green Cadillac with tall, flashy fins. He left the household a month before Pearl gave birth to William and moved in with a woman half a mile away, with whom he had a daughter. When Outlaw was in grade school, his father would occasionally announce that he was going to take him fishing, and Outlaw would spend hours on the stoop at Q View patiently waiting for his father, who would never show up.

Outlaw's father was not a criminal but it was he who visited the most violence on the household. Pearl would call her ex-husband whenever she felt she could no longer control her youngest son, either because he was skipping school or getting into too many fights. Outlaw's father would appear promptly, bursting into the apartment, and overpower his son. He would pull down his son's pants and beat his behind with a belt. Often all it took was ten vicious whacks to produce the desired amount of blood and welts, and a burning sensation that his son would not soon forget. Glancing up at his father during the beatings, Outlaw would see his frenzied face, looking deranged, yet at the same time almost joyful. Even at his young age, Outlaw sensed that his father's rage was somehow misplaced. Sure, Outlaw had done something wrong; sometimes Outlaw himself believed that he deserved punishment. But he could feel in his heart there was no love behind his father's so-called discipline.

The violent behavior extended beyond the nuclear family. Outlaw's older cousin James also grew up at Q View, and was aggressive and beyond control from an early age. Outlaw didn't see a lot of James when they were growing up because James was always "away" at some youth correctional facility or other. As an adult, James would be sentenced for first-degree robbery and would be at the center of a riot involving seventeen inmates. Additional time was tacked on to James's sentence because he nearly killed his stepfather, a fellow inmate, with two homemade knives. He contended the stabbing was justified because his stepfather had murdered James's mother.

When Outlaw was six years old, his older cousin Rufus, who was in his twenties, came to visit. Outlaw had only met him once before, at some long-ago family cookout. Rufus arrived in the middle of the night. Outlaw awoke to see Pearl tending to Rufus, who was lying on the couch and bleeding profusely from his midsection. Pearl shouted at Outlaw to go back to bed. Through his bedroom door, Outlaw heard the hushed activities the rest of the night, the cries and the whimpers, the sound of his mother attempting to nurse her nephew. Outlaw tried to rid his mind of the sight of Rufus's blood oozing from around his stomach. For the next three weeks, Rufus lived on Pearl's couch, becoming more alert and mobile with each day. Pearl, ever industrious, nursed him well and her patient was young and strong. Outlaw never once heard his mother complain about this extraordinary imposition; to Pearl, apparently, such interruptions in routine were just something you did for family when you were called upon. Neither Pearl nor Rufus ever shared with Outlaw how or why Rufus was shot, but Outlaw somehow intuited that his cousin did not have the option of having his wounds treated at the hospital because doing so would attract the unwanted attention of the authorities. One day, just as abruptly as he had arrived, Rufus disappeared and Outlaw never saw him again. Years later, Outlaw heard a rumor that Rufus had died of AIDS.

Pearl Outlaw, in her propriety, saw herself as fiercely at odds with the men of her family's intergenerational cycle of violence. Her correctives were constant work, her famous soul food, and her ever-growing faith, which she expressed at the Mount Olive Baptist Church. Pearl too had grown up without a father, with her sisters and mother in Oak City, North Carolina, population 500. In her work ethic, Pearl took after her mother, who had retreated to a patch of land where she raised cattle and pigs and grew okra and corn, with a shotgun at the ready at her front door. Outlaw's grandmother had a well-justified dislike of white people. She'd done agricultural piecework for white farmers all her life, and often been shortchanged. She greeted anyone—black or white—who drove up to her house with a cocked gun aimed at their heads.

Pearl's kitchen was filled with sumptuous-smelling collard greens, fried chicken, and okra. The bills were paid with utter scrupulousness. But there was not room for much else: no eating out at restaurants and no family vacations, ever. One Christmas, Outlaw prayed for a tricycle, making his wishes known to Santa, but the gift never arrived. Pearl's one indulgence was a shiny red Buick, which everyone called "Miss Pearl's car." Pearl washed the car every weekend, polishing its white vinyl seats.

On her rare nights off, Pearl and Outlaw watched television— *Monday Night Football*, *Laverne & Shirley*, *Happy Days*, and boxing matches on ABC with Howard Cosell. Pearl always called Outlaw by his middle name: Juneboy. After all, he was her youngest, her baby. Whatever havoc he might be causing elsewhere, around his mother Outlaw was always gentle and respectful. Next to her on the couch, Outlaw would notice a thin scar that ran down the entirety of his mother's face, from her eyebrow to her lip. He wondered about its origin, but never dared to ask.

IN THE MONTHS that followed the robbery with Stevie, Outlaw didn't reflect much on the incident. As far as he was concerned, the whole thing was over. The money was long gone, smoked up in marijuana and a new pair of Adidas.

But then word spread like a fever through Newhallville that Stevie had been arrested for homicide. At first, no one believed the story, but it turned out to be true. He had killed his girlfriend's father in the living room of his girlfriend's house. The father, presumably unhappy with his daughter's choice of partner, had confronted Stevie and in retaliation, Stevie had fired upon him. The shooting had occurred on Valentine's Day. The murder was the talk of Newhallville for weeks, but Outlaw shrugged it off. He found the charge oddly relieving: surely no one would ever care now about that other crime, the petty thievery near Visels Pharmacy.

But Outlaw came home one day to find Pearl seething.

"Did you do that robbery, Juneboy?" she said.

She explained that a juvenile probation officer had just visited her, and he had said that a nice man had reported being robbed at gunpoint near Visels by a boy matching Outlaw's description.

It never occurred to Outlaw to lie. He would never dare do that to his mother.

"Yes, Ma," he said.

"Well, you better tell that probation officer exactly what you did."

The next day Outlaw and Pearl went downtown to the courthouse and he was placed in a lineup of young black boys. All the boys were dismissed but William. Some kind of hearing then transpired. Outlaw was given the opportunity for defense counsel but Pearl said, no, that would not be necessary. Outlaw was charged with assault with a dangerous weapon as a juvenile, and sentenced to four months in a juvenile detention facility, where he could finish out eighth grade.

Pearl was relieved. She had been feeling for some time that she could no longer control her son, and she was happy that he was under the temporary control of the state. Outlaw didn't care much either way. He knew the sentence wouldn't stick, and wouldn't be part of his permanent record. That's what all the kids around Newhallville said. But the officials seemed to take the charge very seriously. Some judge or other at the sentencing looked sternly at Outlaw and implored him to straighten up his life. It wasn't too late, the man said. It took all of Outlaw's self-control not to laugh. The judge also explained that Outlaw wouldn't be returning to Shelton Avenue, but going directly to a detention center until a bed opened up at the juvenile home in Middletown, twenty-five miles north.

For this at least, Outlaw was grateful. This way his father couldn't get to him. Nothing that the law could dole out would be any worse than the beating his father would have delivered.

The Connecticut Industrial School for Friendless Girls in Middletown—the site of what would turn out to be half a lifetime of institutional confinements for Outlaw—opened on June 20, 1870. In the original planning documents, the "proper subjects" for care or institutionalization at the school were described as "1) The stubborn and unruly . . . 2) Truants, vagrants and beggars . . . 3) Those found in circumstances of manifest danger of falling into habits of vice and immorality . . ." and finally, "4) Those who have committed any offence [sic] punishable by fine or imprisonment, or both, other than imprisonment for life." But by the time that Outlaw arrived at what was now the Long Lane School in 1983, transported by a probation vehicle, the facility served only boys in trouble with the law.

Pearl did not accompany Outlaw on the trip, nor did she visit much during Outlaw's time at the facility, subscribing to the theory that some "tough love" toward her son was overdue. Long Lane occupied a three-story, sprawling building with a rather noble white cupola at its peak, intended, it seemed, to give a measure of grace to the place. The facility was surrounded by thirty acres of rolling meadows. All of these soothing touches, however, were undermined by the interior of the facility, which was dark, creaky, Gothic, and poorly maintained. Upon arrival, Outlaw was fingerprinted and underwent a series of psychological, medical, and intellectual assessments. He dutifully signed the forms, and was placed in a four-man dormitory room.

He found himself surrounded by earnest young counselors and teachers. They were overwhelmingly white, recent graduates from state universities and liberal arts colleges. Outlaw steadfastly declined their overtures to discuss his "issues." It wasn't that he disliked any of them; he simply had nothing to say to them. They had no comprehension of the world he came from. But Outlaw was compliant enough, and was quickly moved from the central building to the "cottages," homier buildings scattered throughout the grounds. Unlike the main building, the cottages were not secured: while a curfew was in place, there was nothing to stop Outlaw or anybody else from escaping at any time. One day, a group of boys said they were bolting, and asked Outlaw to join them. "No thanks," Outlaw said. He saw no reason to leave. He liked the gym at Long Lane, and had recently discovered a pool, which the nicer counselors allowed him to swim in during the evenings. He had never been at a place with such amenities.

Mainly he played basketball at the facility. On the playgrounds of New Haven, he'd evinced an early talent for the game with his unique combination of strength and size and quickness. There was even talk that Outlaw might be among the first out of the New Haven projects to make it to the pros, either in basketball or football. Outlaw knew that was just the fantasies of schoolkids. But at Long Lane, with virtually no distractions other than some desultory classroom work, he played basketball for hours at a time with a relentless hunger to improve. The other boys made fun of him for his single-minded focus, but he didn't care. He got quicker and stronger, and realized, with a kind of inner joy, that he wanted to be great at something. Maybe it wasn't even going to be basketball; he just loved the idea of applying himself fully, of cutting no corners. The other boys talked incessantly of their plans when they got out, which mainly involved stealing cars. Outlaw was bored by these conversations, and now realized that if he was going to continue in the criminal game, he wanted to do something major: make some real money, and create some kind of organization. He was beginning to see that he might be able to live on a larger scale. After four months at Long Lane, Outlaw was told that he had successfully completed both the eighth grade and the terms of his sentence, and was congratulated by all the well-meaning counselors. Outlaw thanked them all graciously, and kept it to himself that he couldn't wait to get back to New Haven and start his criminal career in earnest.

OUTLAW WAS DELIVERED to Pearl's townhouse via a juvenile probation officer's Ford Crown Victoria, and greeted perfunctorily by his brother and sister. Pearl made a cake, and the family celebrated his return with aunts and uncles and cousins. But after the guests had departed, and Pearl and Outlaw were alone, she began her lecture. She was happy to see him home, she said, but now was the time for him to clean himself up. "I don't work so hard for my son to be a thug," she said. "What am I doing all this work for?" Outlaw nodded in agreement. He loved his mother and didn't want to upset her.

The next day, when Outlaw walked onto Dixwell Avenue, he was greeted uproariously by his peers. Outlaw was now fourteen—although he could pass for twenty—and was fit and muscular from all the gym work while he was away. He had a newfound imposing, almost smoldering look, and was welcomed as if he were some kind of conquering hero. People looked at him differently, with more "respect." The fact that he had now done significant institutional time, and had been an associate of Stevie, a known murderer, drastically elevated his street cred. Outlaw sensed that he might be able to profit from his new image.

Still, with Pearl's speech in his mind, Outlaw wanted to explore a legitimate opportunity that he had just heard about. The kids were talking up a new government summer program for underprivileged youth to clean city parks. Outlaw walked two miles across town to the Parks and Recreation Department, where he was handed an application amid a crowd of candidates. He filled out the form slowly, in the large cursive handwriting that he had been taught in school, and was told to sit in the waiting room. After some time, his name was called. A clerk told Outlaw that he didn't qualify for the job. "Your mother makes too much money," he said. Outlaw felt like crying the whole way back to Newhallville.

A week later, Outlaw was hanging out on the front stoop of the townhouse. There had not been much to do since he'd been rejected for the job. He'd watched a lot of reruns on television. An older man, Ducky, came up to him. Outlaw knew Ducky by his street reputation, as a low-level pot dealer.

"Hey, you want to deal some pot for me?" Ducky said.

Outlaw may have hesitated for a moment, but it wasn't for long.

"Yes," Outlaw said. "I'll deal for you."

Ducky brought him to Lilac Street and told him how things would work. Outlaw would sell individually packaged wax paper bags of marijuana for $5 each. For every $100 of product that Outlaw sold, Ducky would get $75 and Outlaw $25. When Outlaw ran out of bags, Ducky, who would always be circling the block, checking on things, would resupply him.

Outlaw dealt for Ducky the next day and made $200. For the remainder of the summer, he averaged $300 a day, and on Friday and Saturday nights up to $500. He bought plenty of pot for himself, and lots of new clothes, the latest hip-hop fashions from New York. For all of Outlaw's difficulties with keeping track of words and figures in school, he was excellent at the business of drug dealing. Ducky only had to tell him something once for Outlaw to get it immediately. In back alleys, Outlaw made the drug and cash exchanges quickly and with total discretion, to such a degree that he was not once accosted by the police all summer. He knew how to keep things moving "in and out," as expeditiously as possible. At the same time he was affable, even jovial, with the clientele, treating them with respect regardless of their age and color, and showing no judgment toward the hard-core users who were a significant portion of his customer base. Outlaw understood that addiction was just part of the landscape of New Haven, or at least his New Haven, just the way trees or birds or cars were. Ducky told him he was "a natural-born hustler."

Surely, Pearl must have suspected that Outlaw had entered a new profession. There was plenty of evidence—all the new clothes and her son's bloodshot eyes when he came home at night—but she said nothing. William's father too wasn't coming around much anymore, perhaps realizing that his youngest son was now too big for him to deliver a beating without fear of retaliation.

In September, almost as an afterthought, Outlaw began high school at Wilbur Cross, a mile away from home. Cross, as it was called, teemed with thousands of students. From the beginning, Outlaw's intention for his academic career was for it not to interfere with his drug dealing. His main interest in high school was Cross's nationally famous basketball program, but that wouldn't start until November. Outlaw woke daily at five, trying to get a lucrative hour of dealing on Lilac Street before the first bell at school, and then he would return to sell more in the afternoons and evenings. Many days he skipped school entirely. Given Outlaw's juvenile record, school officials at Wilbur Cross were concerned about him even before he arrived. They dispatched a young Irish-American police officer, Edward Kendall, to meet with him. Kendall had been on the force for four years and was known for his tough but fair style, and his ability to connect with kids. Outlaw and Kendall met in a guidance counselor's office, the first of what would be a lifetime of such encounters. "I've seen you run down the streets of Newhallville. Probably from the police!" Kendall joked. "You run like a deer. With your size, you should be playing football. I bet you could make the NFL. Seriously." Outlaw shrugged. He wanted to play basketball and didn't understand why Kendall was bothering to talk to him. Didn't the police have better things to do? Kendall gave Outlaw his card, and told him to call anytime, for any reason. Outlaw threw the card away.

Soon enough, everyone in school knew Juneboy. Previously he had been perceived as a big lug, but now he was a known dealer with flashy clothes. One girl in particular intrigued him. Her name was Phaedra and she was a celebrity around Wilbur Cross, a flashy hip-hop dancer in a troupe. She was beautiful and her smile lit up the stage. Their relationship began with little teasing asides in the hallways. Always flush with cash, Outlaw took Phaedra out to fine restaurants and brought her flowers. They began spending hours together in Phaedra's bedroom. Her mother caught them repeatedly and locked Outlaw out of the house. But he would crawl back in through a side window.

Sometimes instead of giving Outlaw pot, Ducky would hire him by the shift. Outlaw did this for a few weekends, but after failing to get paid, Outlaw calculated that Ducky owed him $1,500. Whenever he saw Ducky around, he'd ask for the money but Ducky always had some excuse. Outlaw decided he needed to buy a gun to protect himself, a .25-caliber revolver. Acquiring it was as easy as buying a loaf of bread on Dixwell Avenue. Outlaw bought the piece from a junkie for $80. Upset at Ducky, Outlaw was hanging out late one night on Lilac Street when an older dealer named Butchy, a slick guy with a goatee and a smooth black leather jacket, pulled over in a car. Butchy, unlike Ducky, was a big-time dealer, considered "the Godfather" of Lilac Street.

"I'll offer you seventy/thirty. I'll do better than Ducky. He's squeezing you," Butchy said.

Outlaw hesitated. He didn't really like Butchy and was getting tired of working for people. But then he saw Butchy's car, a sleek gray Porsche. Outlaw had just watched *Scarface* with Al Pacino, and he realized that Butchy was driving the same car that Pacino had in the movie.

"Sure," Outlaw said. "I'll deal for you."

Under Butchy's tutelage, and now armed with the proper equipment, Outlaw began to earn consistently $500 a day. He was climbing the ranks of his newfound profession.

Just before Thanksgiving, basketball season began, and Outlaw finally had a reason to attend high school. The force behind Cross basketball was a tough and charismatic coach named Bob Saulsbury. Saulsbury's teams were nationally ranked during the 1970s and a recent Cross graduate, John Williamson, was Rookie of the Year for the New York Nets. Saulsbury considered himself a disciplinarian whose dedication to the clean habits of his players was a matter of life and death. Outlaw bypassed the JV entirely and began playing meaningful minutes as a power forward for the varsity team, highly unusual for a freshman. Beyond his size, he had a natural sense of the game, an unteachable sense of timing and space. During a scrimmage in which the second team played against the varsity, Outlaw dominated against the senior starter in his position. In the showers afterward, Outlaw's humiliated rival slapped him, clearly itching for a fight. Outlaw, calm as ever during conflict, pummeled him badly, leaving him bloody and bruised on the tile floor. Some of Outlaw's teammates congratulated him afterward.

For a freshman, Outlaw had a great season, averaging ten points a game, but in the stretch run of the state basketball playoffs, Saulsbury benched him. The coach had heard rumors from the other players that Outlaw was running in the streets, and could see Outlaw's lifestyle reflected in his flashy clothes. Outlaw watched from the bench as the team went on to easily win the state championship. The day after the season ended, Outlaw stopped going to high school. A week later school authorities called his mother, and thereafter began a long series of meetings and "case conferences" involving assistant principals and counselors and social workers. Perhaps Outlaw wanted to take a break, then go back to school in a month or two? Or attend summer school? Or receive home tutoring until he got back into the swing of things? Outlaw said yes to all of these in the presence of his mother, but skipped all of the carefully arranged appointments. Lilac Street was much more interesting.

A month after Outlaw's last day at Wilbur Cross, Pearl found the gun under his bed, along with $3,000. She said, "You can't live in my house and live that life."

Outlaw knew there was no point arguing with Pearl. He went upstairs and packed a bag.

"You're right, Ma. You're absolutely right," he said to her.

Pearl cried as he walked out the front door.

CHAPTER 2

The Jungle Boys

Sixteen-year-old William Outlaw.

THE HOUSING AUTHORITY OF THE CITY OF NEW HAVEN WAS ESTAB-
lished in 1938, only one year after Congress passed the Housing Act, which
allowed for the creation of permanent public housing for low-income residents in cities across the United States. The initial motivations behind the
new institution of public housing, both in New Haven and nationally,
were arguably well intentioned. After the First World War, throughout
the country, but particularly on the East Coast, immigrant families and
former agricultural workers flooded the cities seeking factory work. Many
of these newly arrived families lived in cramped conditions—a family of
eight or ten living in a single room was not uncommon—in run-down
tenement buildings often teeming with lice and vermin and infectious

diseases. The new legislation attempted to redress those conditions and replace the tenements with clean, new, standardized, often high-rise apartments, with plenty of open space and fresh air between the buildings.

However, the framers of the public housing movement made a critical error as they conceived of their master plan: they prohibited tenants from owning their apartments, and therefore residents had little personal investment in the properties in which they lived. The buildings were owned by city authorities and often poorly maintained. The projects were frequently opened with great optimism and fanfare, but then quickly fell apart. And just as often, in creating the new projects, entire neighborhoods—many of them containing handsome historical structures alongside the squalid tenements—were razed. These processes created deep-seated feelings of dislocation and alienation among long-standing residents. In New Haven alone, 20 percent of the population was forced to move out of their homes between 1956 and 1974, all in the name of purported urban renewal and progress.

Whether intentional or not, the vast majority of public housing served African Americans. An official policy of the Federal Housing Administration between 1934 and 1966 severely curtailed homeownership opportunities for minorities. Informally called "redlining," the policy graded neighborhoods in cities across the country into four levels, based on desirability and the quality of housing stock. The worst, or "fourth grade," level, often where many African Americans lived, were colored red on government maps. In one representative example, the red-line area of a city was characterized as having "*detrimental influences* in a pronounced degree, [an] *undesirable population or infiltration of it.* Low percentage of home ownership, very poor maintenance and often vandalism prevail. Unstable incomes of the people and difficult collections are usually prevalent. The areas are broader than the so-called slum districts." The government advised mortgage lenders to be conservative in making loans for housing purchases in red-line districts, essentially freezing out minorities and the poor from opportunities for private ownership. There was no place to live but the projects. In New Haven, the red-line districts included the areas around Q View, Newhallville, and the Hill—in other words, William Outlaw's neighborhoods.

For sixty years, New Haven's leaders had anxiously targeted one particular area for reform. This was the Church Street South neighborhood across from the train station, which for much of the nineteenth and twentieth centuries was a warehouse area and open-air wholesale market, surrounded by rickety tenement buildings. Politicians were particularly vexed by the area's high visibility: the rough-and-tumble market was the first thing that new arrivals by train saw as they entered the city. Richard Lee, the reform-minded mayor of New Haven between 1954 and 1970, described the neighborhood as "a tangle of stress, often so congested that normal business was impossible. Most business was conducted from the tailgates of trucks . . . the streets were too often littered with refuse and filth and infested with rats and vermin. . . . One can hardly imagine a less impressive entrance to a city."

Urban planners had long fantasized about grandiose schemes to permanently remove this stain from the city's canvas. Indeed, the location of Church Street South—only a few blocks from Long Island Sound, and a third of a mile from the stately New Haven Green and the city's growing financial district—held great promise for development. In 1910, the city hired Cass Gilbert, one of the country's preeminent architects, and Frederick Law Olmstead, Jr., one of its most established landscape architects, to create a civic improvement plan for Church Street South. Olmstead and Gilbert proposed that a Parisian-style boulevard be constructed on the site, linking the train station to downtown. The plan was rejected by the mayor, who found it impractical and expensive. Thirty years later, a planner envisioned the Church Street South tract as a commercial paradise, à la Manhattan's Fifth Avenue. Nothing was done with that idea either. The truck market stubbornly endured for another two decades. In 1965, Mayor Lee proposed that Church Street South be transformed into a luxury housing complex with an expansive park and school and a boy's social club to be designed by the world-famous German-American architect Mies van der Rohe. But the seventy-nine-year-old van der Rohe, in poor health, became too frail to work on the project and dropped out. Finally, amid the Civil Rights Movement and a growing realization in the late 1960s that New Haven lacked adequate affordable housing, the plan switched once again, this time to a much more modest proposal, a low-income-housing project. But this public housing endeavor, it was promised, would be different. It wouldn't be like Q View or the other disastrous projects around town. While the project wouldn't be the creation of a world-famous architect, it would be designed by the dean of the Yale School of Architecture, Charles Moore, and would feature inspiring, contemporary styles. The idea was to create not a high-rise as originally intended but a kind of modern Italian hill village in earthy, pastel colors. Moore conceived of a series of three- and four-story concrete duplexes linked by pedestrian paths and small plazas. The project would be ambitious—three hundred apartments in all, some of them with up to five bedrooms. Under Mayor Lee's steady hand, the complex, still called Church Street South, was completed rapidly with federal funds and opened in 1967. The project won an architectural award shortly after opening for its bold and pioneering vision.

SIXTEEN YEARS LATER, William Outlaw, recently kicked out of the townhouse on Shelton Avenue by Pearl, knew exactly where he was going to move: Church Street South. Outlaw was unaware of the long history of the area and cared even less. All he knew was that Charles Moore, in his high-minded schemes, had inadvertently created the perfect environment to deal drugs. In the previous decade and a half, the Tuscan-inspired winding pedestrian pathways had become drug hot spots that the police could barely see, let alone access. Church Street South had actually deteriorated far more quickly than other housing projects around town. Moore's plan had called for flat roofs for the apartment buildings, an absurd choice in New England. In a bad winter, feet of snow would linger on the roofs until springtime, and then slowly melt, the water leaking into the heated apartments and creating a black mold. Many children growing up in the complex developed chronic asthma.

Just a block away from Church Street South was New Haven's bunkerlike, almost windowless police station, another Modernist concoction. Outlaw considered the proximity of the police station an advantage. With his deep-seated hatred of the police emanating from Q View, he planned to monitor them as much as they watched him. The final attraction for Outlaw was Church Street South's closeness to the train station: only fifty feet away, it ran commuter trains every half hour to New York City, providing both a steady supply of customers and a direct conduit to Harlem and the Bronx, the source of most of the drugs in New Haven.

The same day that he left his mother's apartment, fourteen-year-old William Outlaw, resembling a grown man despite his age, stood in the middle of Charles Moore's dense complex, at the top of an oval-shaped area that in later years Outlaw and his crew would call "the hill," though it barely merited that term. The rise was only ten feet from top to bottom over a span of thirty yards. At the top of the hill was a copse of sickly pine trees that looked as if they'd been poisoned, which they probably had been by the decades-long detritus of toxins from the truck market. Outlaw considered the possibilities that lay before him. One might think that he would be anxious. He didn't know where he would sleep that night, and it had only been a month since he dropped out of high school. On the other hand, he had grounds for optimism: he reassuringly had a gun in one pocket, a roll of thousands of dollars in the other, and a steady means of income from his ongoing weed business on Lilac Street.

OUTLAW KNOCKED ON the door of apartment 12. It was beige, in keeping with Charles Moore's Italianate vision. The rap of Outlaw's fist on the door made a thin and hollow sound. Outlaw knew who lived in apartment 12, and he wanted to talk to her. Her name was Sheila. He didn't know exactly how he knew Sheila; he just knew her in the way that he knew many people in New Haven, particularly black people who were around his age. Sheila was only five years older than he was, a teacher's aide and a single mother with a baby. Most important for his current purposes, Outlaw knew Sheila to be a cocaine addict, and figured she wouldn't mind having an extra few hundred dollars a week.

"Hello, Juneboy," Sheila said, opening her door. She seemed unsurprised to see him, or anybody else, for that matter.

Outlaw got right to the point, which, as a new businessman, he was learning was the best way to get things done. It was March and the weather was raw, and he would prefer to get inside as soon as possible. He asked if he could stay in her extra bedroom for $200 or $300 a week, cash. He would help out with her kid sometimes if she wanted. And depending on how things went, he might be able to help her out with a new TV, maybe some new furniture.

"Sure," Sheila said. She ushered him in. Her apartment was clean and smelled of soul food. The baby girl's name was Kitten. Outlaw stayed in Sheila's spare bedroom that night, played with Kitten on his knee, and slept in while Sheila went off early the next morning to work. Later that day, while Pearl was at the gun factory, Outlaw picked up his clothes from Shelton Avenue and brought them to Sheila's place. Outlaw didn't tell Pearl where he had moved, but it didn't take long for her to figure it out through the grapevine. She charged down to the police station with a picture of her son and informed as many officers as would listen that he was living at Church Street South, that he was a minor, and if they saw him, they had better arrest him for trespassing. After that, she marched into every bar in Newhallville and around Church Street South and told the bartenders that if they dared serve her son a drink, she would sue them and have their liquor licenses removed for serving underage customers. (Pearl's campaign worked brilliantly in one aspect: for the next five years, even though he went on to become New Haven's biggest drug dealer, Outlaw was never served a single drink in the bars that Pearl had targeted.)

Over the next few months, Outlaw slept at Sheila's and commuted to Lilac Street. He would leave early in the morning, walk the two miles to Newhallville, sell pot for Butchy, or now sometimes on his own, for fourteen hours, and walk the two miles back to Sheila's late at night. If Sheila was still up, they'd watch TV together. Even though she was only a few years older than he was, there was something rather maternal in her relationship with him. She'd leave out okra, fried chicken, and hamburgers for his dinner. With his profits from Lilac Street, he paid his rent weekly as promised, beginning a lifelong habit of prompt payments. Sheila upgraded the appliances around the house, and also used the funds to support her snorting habit. Outlaw began to like his new life. He missed his friends from high school, and of course Pearl, but he was very happy dealing pot.

When Outlaw left Sheila's place every morning, he often noticed a commotion around one apartment. A woman lived there, and she had all kinds of people coming to her door and kitchen window. She was in her thirties, with an eye-catching figure. He asked Sheila who the woman was and she said her name was Suzie Jones. One morning Outlaw delayed his walk to Newhallville and in the chill April air observed Suzie from across the courtyard. Over the course of an hour, he watched while numerous customers, many manifesting signs of addiction—including spindly legs, sunken eyes, and jittery hands—exchanged twenty-dollar bills for little plastic bags from Suzie. Outlaw knew that the drug being traded was cocaine, but he was not particularly familiar with it. He didn't really know how cocaine was measured and sold, but he noticed a zeal around these exchanges that was unlike anything that surrounded the sale of pot. The customers wanted the stuff—*needed* the stuff—with a frenzy that Outlaw had never seen before. Outlaw sensed an unprecedented business opportunity.

WITH HIS NEWFOUND freedom as a high school dropout, Outlaw occasionally took the day off, walking across the street and hopping the train for New York City. Typically, he went to Harlem. He loved the energy of 125th Street: the clubs, the women, the street hustlers, the pageantry. The city perfectly matched his size and personality. Sometimes he'd go to Midtown, where he bought the latest sneakers and hip-hop gear in styles that had not yet reached New Haven. One day he found himself on West Forty-seventh Street and happened upon a block filled entirely with jewelry stores. This was the world-famous Diamond District, through which 90 percent of the gems in the United States were packaged and processed. Outlaw, with his steady supply of cash—he had several thousand in his pocket that day—wanted a rope: a thick gold chain necklace that would announce his arrival as a bona fide drug dealer. In his new world, ropes were tantamount to a business card, identifying their wearer as a person of stature on the street. Entering one of the hundred jewelry stores on the block, he found what he was seeking: a thick chain-link gold necklace, exuding masculine power. Outlaw purchased the chain for three thousand dollars. Immediately after the transaction, a man with a strange-sounding accent approached him.

"Anything else I can help you with?"

Unsure of the man's intentions, Outlaw scowled at him. But the man gestured to Outlaw to follow him to the back of the store, and Outlaw warily did so. Discreetly, the man handed him a small baggie containing a hunk of a whitish substance. It looked like a piece of chalk except for an iridescent sheen on its surface. "Take this," the man said, "for free. If you like it, come back, we have more for you. If you don't like it, just keep it. No problem." Outlaw put the bag in his pocket. On his return to New Haven, he went to Lilac Street and showed the bag and its contents to a man whom he knew to be a hard-core addict.

"Shit!" the man said. "Let me see those fish scales. That's fucking pure cocaine! See how shiny it is? Right off the kilo. That's a beautiful thing, man. Where the fuck did you get that?"

"Just got it." Outlaw shrugged. "I want you to try it. For free."

"Fuck yeah! I'm going to turn this into crack!"

With a penknife Outlaw shaved off a chunk of the fish scales into the man's shaking hands.

"Th—thh—thank you," the man said.

"But you got to come back and tell me how good this shit is," Outlaw said. "I want the truth."

The man returned half an hour later, barely able to talk. He was so stoned that his lips could hardly move and he had difficulty forming words.

"J—J—Jjjuneboy. That's s—s—some good shit. Get me some more, okay? Please?"

LATER THAT NIGHT, after Kitten went to bed, Outlaw cut off another chunk from the fish scales and gave it to Sheila. Outlaw watched as she took a razor and transformed the piece into a fine powder, then snorted it with a dollar bill. Sheila paused, waiting for the drug to take effect. Turning her head upward with pleasure, she said, "Juneboy, you gonna be a millionaire! That's the best cocaine I've ever had."

Outlaw asked Sheila, in her delirium, "How much does a kilo cost, Sheila?"

"I don't know, twenty to twenty-two thousand? When you break it down with lactose powder, you can make six or seven times what you paid for it. But how you gonna get a whole kilo of it?"

For the next two weeks, Outlaw worked morning to night on Lilac Street selling as much weed as he could. Then, with nine thousand dollars in his pocket, he hopped the train and returned to the Diamond District.

"Oh, you're back," said the man in the back of the store, unsurprised. "You interested in some more?"

Outlaw nodded.

"Can you take a ride with me? Don't worry, it will be okay," the man said in his stilted accent. He wore a collared shirt and pressed suit, and had his dark hair slicked back. Outlaw thought he resembled a mafioso, but a classy-looking one, not like the wannabes in New Haven who looked like they had stepped out of a bad Frank Sinatra movie. The man made a call and within minutes a car and driver pulled up outside. The man opened the backseat passenger door of the large black sedan for Outlaw. Normally, Outlaw would never take such risks, but he heard Sheila's words—"You can make millions off this shit!"—echoing in his head. Outlaw spread out in the backseat, but not before feeling for the revolver in the holster under his belt. The man from the store joined the driver in the front, who had a similar accent. Within minutes the car pulled out of Manhattan and across the Fifty-Ninth Street Bridge, the East River a hundred feet below them.

"Where the fuck you guys from anyway?" Outlaw said.

"Albania," the two men said.

Outlaw didn't know where Albania was. In the end, he decided, it didn't matter.

"Well, if you're from Albania, I'm from Albania too," Outlaw said.

They all laughed in unison.

The men tried to ask Outlaw questions about himself, but he parried them all. He didn't want them to know where he lived. Within twenty minutes, they pulled into the driveway of a two-story townhouse in a respectable middle-class neighborhood somewhere in Queens. From the constant drone of planes overhead, Outlaw thought they were near LaGuardia Airport.

"We go inside," the men said. Outlaw knew that anything could happen inside that house, but he somehow felt that things were going to work out. There was something—what was the word?—*professional* about these guys. Something matter-of-fact, appealingly straightforward. And, of course, he had his gun. They walked up the stairs to the second floor, where there was a modern open kitchen and a living room filled with plush furniture and a glass table. On the table were ten blocks of a yellowish-white chalky substance, each the size of a brick and stamped with the letters *PAN.* In time Outlaw learned this was short for "Panama": the cocaine had likely been produced in Colombia but had passed through Panama with the cooperation of General Noriega's government.

The Albanians asked him to sit down and inquired what he wanted to drink. Coffee, vodka, soda, whatever. "No thanks," Outlaw said.

"How much do you want to buy?" the Albanians said.

Outlaw said he had $9,000. He wanted however much that would buy.

"Well, we sell it for nineteen thousand a kilo, so we'll give you half a kilo. Discount for first-time customer!" They laughed.

Nineteen K a kilo? Outlaw thought. *That's fucking great. Sheila said it would be 20 to 22K.*

THE NEXT DAY, with the help of two older drug dealers from Newhallville who operated both independently and for people like Butchy, Outlaw spent an entire day breaking down the half kilo in Sheila's apartment, transforming it into the finest powder they could. To the cocaine they added lactose, a powder made from dried milk. The resulting mixture was 60 percent coke and 40 percent milk powder. With a scale, they measured out seven-tenths-of-a-gram amounts that they then put into small wax paper bags to be sold for $20 each. Outlaw and his men then placed fifty of the twenty-dollar baggies into one large plastic bag. With a kilo, Outlaw estimated, they would be able to make seventy such large bags. In this way, a $19K purchase of a kilo in Queens could be converted to $70,000 on the streets of New Haven.

On the periphery of Church Street South, Outlaw and the two men set up shop by a dumpster. They sold the bags in quick transactions out of view of the roadway. A number of customers returned for more, saying it was the best cocaine New Haven had ever had. Within a few days, the half-kilo was sold out. Outlaw went back to New York, bought a full kilo this time, sold that in a week, and repeated the process. Within a month he was buying two kilos at a time, and recruiting more men to help, which, given the thousands of dollars Outlaw could now pay them, wasn't difficult. Thanks to the Albanians—or whatever their nationality—even Outlaw was surprised at how quickly he had become a legitimate cocaine dealer, even if he had since learned that he hated the drug himself. He sniffed the powder on a couple of occasions but despised the wired and edgy feeling it left him with. He was alert enough on his own.

In those first few energetic months, things could not have been going better, but Outlaw knew he had a major problem. The spot near the sidewalk where he was selling was terribly exposed. The police could bust him at any time. Outlaw needed to get his operation inside Church Street South, where Suzie Jones was. The answer was simple: he had to get her out.

In observing Suzie, Outlaw noticed how lean and jumpy she was, and that her complexion had a gray, unhealthy pallor. He was sure she was a heavy user in addition to being a dealer. Outlaw also saw how the police were onto her. The cops knew that she was dealing, but they couldn't just barge into her apartment without a warrant, so they had developed an alternative, if illegal, strategy. Lying around Church Street South were dozens of cinder blocks, dismal remnants of the shoddy construction process. The police took to tossing stray cinder blocks through Suzie's kitchen window. In the ensuing chaos, Suzie would flush her product down the toilet and be forced to close shop while waiting for the housing project's handyman to replace her windows. Often this stopped business for days.

One afternoon, Outlaw ambled toward Suzie's apartment, picking up a broken cinder block on his way. He tossed it through her kitchen window. Suzie ran to the window, looking possessed.

"The police were just here," Outlaw said with the calm that he could now expertly summon upon demand. "They tossed the cinder block through the window, then they left. I don't know why they would do that." Suzie anxiously swept up the glass and called the superintendent. Over the next month, Outlaw repeated this procedure two or three more times with Suzie becoming more upset and paranoid with each instance.

And then one day Suzie was just gone. The word was that the police had caught her directly in the act of dealing and packed her and her assistant off to jail. They'd apparently nabbed her with a lot of product and she was going to be away for a while.

This was the moment that Outlaw had been waiting for. He moved the crew inside Suzie's complex and set up shop in the warren-like alleyways surrounding it. Overnight, business doubled. Outlaw hired six more men and went to New York weekly to buy from the Albanians. He spent days cutting up the kilos, processing so much of the drug he often got a contact high from all the cocaine dust in the air. It was now summer, and word had spread throughout New Haven about the quality of his product. Sometimes his men would sell fifty small bags in as little as an hour or two, and Outlaw found himself with as much as $20,000 in his pocket by the end of the day. He couldn't believe how much his life had changed in only a few months. He just didn't know how long it could last.

One June night, Outlaw walked into a movie theater alone. He had no expectations for the movie; he was just looking for something to do on a Friday night. It was an R-rated film, but he was granted admission because he looked like a man. He bought a soda and popcorn and stared at the images on the screen: steam rising from the Manhattan streets, kids fighting with bare knuckles, the killing of a boy by a police officer, the manic selling of booze and marijuana, stacks of hundred-dollar bills, a man getting bludgeoned with a gun and shot through the mouth, men in tuxedos drinking champagne in ballrooms, epic shoot-outs between G-men and gangsters, the door of a gorgeous sedan getting riddled with fifty bullets, beautiful and adoring women, betrayals and murders between former friends. Outlaw went on to see the movie, *Once Upon a Time in America,* fifty times.

Once Upon a Time in America starred Robert De Niro and Joe Pesci and told the story of Jewish kids growing up in poverty during the bootlegging era and creating a gang, only to see their creation implode under a crush of deceit. Sitting in the theater, eyes wide open, Outlaw wanted desperately to be like the boys on the screen. They came from nothing and used their wits to rise above it. They grabbed everything after crawling out of the sewers of New York, and didn't care what methods they used to achieve their dreams. Outlaw found the scenes of mobsters in tuxedos, dancing in glass-roofed ballrooms with sultry women, to be romantic and glorious. He dared to imagine that for himself.

Even though Outlaw was a black man from New Haven, and the boys in the movie were Jews from another era in New York, Outlaw found that *Once Upon a Time in America,* unlike anything he had ever before experienced, described the world that he knew—his raging father and grandfather, Q View, Rufus bleeding from his stomach—and its codes. The film mirrored his feelings about the world: that it was a fallen and broken place; that violence was the baseline of all interactions; that cops were pigs; that actions did not have to have consequences because if you were clever enough, you could figure out how to charm your way out of anything. The escape from the fallen nature of the world was radical ambition and honesty, which could be deployed to create something new and whole. Walking out of the movie theater, into the fresh air of the summer night, Outlaw knew that he was going to do something large and beautiful and exceptional. Yes, it would be risky, but it was the only way to transcend the world into which he was born. *He was going to start a gang.*

Outlaw put out the word that he was looking for a crew, and a large group of applicants materialized immediately. With his direct access to the best drugs, his cash, his profile now in multiple neighborhoods around town, his basketball prowess, and his mannish size, Outlaw was already locally famous in New Haven. The factories had withered away, and there were by the mid-1980s few legitimate employment opportunities for young people without high school diplomas, and no shortage of desperate young men willing to trade their lives for a chance at glory. Outlaw quickly had five members, then ten, then fifteen. He proved an astute judge of character and was able to place the boys and men in the roles for which they were best suited. Outlaw knew intuitively how to create and run a team. He'd been on a championship basketball team after all, where everyone had played his role. Outlaw barked out instructions to his new charges, but the orders were usually delivered with humor, a smile, a friendly dig, in a way that made the members want to be part of the journey.

Outlaw did not know this, but Connecticut in the 1980s was actually the perfect place to form a gang. The idea was counter to Connecticut's sedate and Waspy image, with its Currier and Ives villages and toney suburbs fueled by Wall Street bonuses. But research shows that gangs thrive in areas of wide income disparity, and Connecticut had among the highest income disparity in the United States. Just thirty miles from Church Street South were some of the wealthiest zip codes in the United States, towns like Westport, where Paul Newman, Phil Donahue, and Martha Stewart lived. It also was difficult for Connecticut law enforcement to monitor gang activity in the state, transpiring as it did in relatively spread-out areas as opposed to dense metropolises.

Outlaw knew that if he was going to have a gang, it had to have a proper name. One night he was hanging out with a boy his age, Rodney, by a copse of pine trees. Rodney was short and stout, dark-skinned and imbued with a restless, at times reckless energy. He never stood still. Rodney was now living at Church Street South, in one of its farthest and dingiest reaches, in the back by the police station, dealing pot and cocaine and emerging as Outlaw's friendly rival.

As Outlaw and Rodney talked, it was getting dark, making the pine trees and the walls of the complex seem close and foreboding. Across the way, shadowy figures walked in alleyways. The blackish night felt malevolent.

"You know what this place is? This place is a fucking jungle," Outlaw said.

Rodney laughed. "Yeah, it is."

"And we're the fucking Jungle Boys," Outlaw said.

Rodney laughed manically. "Yeah, we fucking are."

The name caught on immediately. Within weeks and months, everyone called the gang—now, when it was expedient, composed of both Rodney's and Outlaw's crew—the Jungle Boys. A year later, hardly anyone called the housing complex Church Street South anymore. Even the police called it "the Jungle."

SOON ENOUGH, OUTLAW was joined by a twenty-year-old man, recently released from prison. This man, called Evans, was more experienced in the ways of crime than Outlaw, and was instrumental in the continued operations and management of the gang. As Evans and Outlaw oversaw the expansion of the gang to twenty and then thirty members, Outlaw created a business model without even knowing what a business model was. He made it all up in his head, never writing a single thing down.

First Outlaw and Evans created a hierarchy of workers. At the bottom were "lookouts," who were stationed on the periphery of the Jungle and who alerted the gang to the presence of the police and other enemies. Outlaw hired younger boys for these positions, and what he mostly sought was an endless capacity for patience, as most of the job involved standing at a post and waiting, and then waiting some more. Outlaw gave his lookouts whistles and flashlights to use as signals. They didn't usually carry guns. Outlaw regarded the lookouts as his most critical employees, the first line of defense. "Without you, we are nothing," Outlaw said. A good lookout made $1,500 a week.

The next step up was "working the bundle," or selling product. Outlaw wanted "people persons" in this position: individuals who could talk to anybody and be professional with all kinds of customers. Bundle handlers had to be trustworthy, or at least mainly so: a certain level of dipping into the product was to be expected. They also needed to be cool in a pinch: if the police raided, they needed to have the self-possession to drop the drugs immediately down the nearest drain. Bundle handlers made $2,000 to $3,000 a week.

Some bundle handlers carried guns, but many did not. This was the job of the next tier of employee: those working "security," who made up to $3,000 weekly, and roved the premises to defend the territory. Of course, security boys had to be assertive and "jacked up," prepared to fire and be fired upon, but they couldn't be reckless. Violence needed to be judicious. Some of the new Jungle Boys sheepishly told Outlaw they couldn't handle the violence. Outlaw didn't judge their squeamishness, and shifted them into newly formed administrative roles in the budding organization: preparing and packaging the cocaine, counting and storing the money, and payroll.

Each Friday was payday. The crew was paid in envelopes thick with cash. Occasionally an employee might be shorted several hundred dollars for being late or high, but typically Outlaw was generous with his compensation. From the beginning, he understood the value of loyalty. If a customer happened to overpay, he'd let the bundle handlers pocket the extra cash. Once his men spent $100,000 on what turned out to be junk cocaine made of talcum powder. Outlaw shrugged off the mistake. "It could have happened to me. You didn't do nothing wrong," he said.

If Outlaw and Evans and his inner circle knew you and liked you, the hiring process was relatively easy. In a conscious reaction to the earlier era of organized crime in New Haven, the Jungle Boys chose not to replicate the Mafia's initiation rituals. In Outlaw's world, when you were really in with the gang—that is, when you'd proved reliable and tough enough over a period of time—you were simply "down." By being "down," you were eligible for promotion and you had a voice in operations. Members of the Jungle Boys were either "down" or "not down." The "not down" members were short-termers on the periphery, and most of them simply wandered off at a certain point, often in a cocaine haze. Many of the Jungle Boys were lost to addiction. Outlaw considered it just the cost of doing business.

Dealing started at 11 A.M. Outlaw and Evans began the day with a meeting to clarify who was working which shifts, what the police were up to, what weather was expected, possible threats from rivals, and so on. Outlaw laid out the instructions with his characteristic verbal tics: almost every sentence was followed with a "You know what I'm sayin'?" and every male was called a "dude." ("The police are out right now, know what I'm sayin'?" "Dude came up and asked for LSD. I said we don't sell none of that shit.") In the summer, and on weekend nights, business rolled on until 4 A.M. The busiest days were always the first and the sixteenth of the month, when public assistance checks were issued. Operations were conducted six days a week; in deference to Pearl's piety and that of many of the other mothers and grandmothers who had raised the crew, the Jungle Boys closed on Sundays. (Outlaw himself often went to services at any of the twenty African-American churches in town. Older women would approach him afterward and say, "I'm praying for you." Outlaw would smile and respond, "Thanks so much. Keep on praying.") Dealing went on in all weather: blizzards or torrential rain, it made no difference. Outlaw almost preferred severe weather, regarding it as a test of his dedication. Only real hustlers sold in a blizzard. Every other dealer in town would close, and all the customers would come to him. "Hustlers got to hustle," Outlaw was fond of saying.

It was the nature of addiction that the business never stopped, except for one week during the gang's first year, when all of New Haven ran out of cocaine. The supply from New York had gone completely cold. "You gotta get me some!" the hard-core users demanded. Some screamed for a fix, while others became so depressed and fatigued from the sudden withdrawal that they could only mumble. As the freeze went on, signs of psychosis emerged: the heaviest users scanned the sidewalk for any sign of drugs, even crawling around on their hands and knees to see if they could find any. Outlaw vowed that such a break in business would never happen again, believing, just like a good bank or insurance company, that the integrity of his operation was paramount. Afterward, if the Jungle Boys bought fifteen kilos, they sold ten and stashed the extra supply in various apartments around town for future use.

Most of his clientele were black, and the remainder a more or less even

mix between Hispanic and white. Outlaw decreed that there be two separate lines for customers: one for the white customers, and one for everyone
else. His reasoning wasn't racist, but pragmatic. Since there were so few
blacks in the police department, there was little chance that a black customer could be an undercover officer, but a decent possibility that a white
customer might be. The bundle handlers in the line for the whites had to
take more precautions.

When members of the Jungle Boys inevitably got nabbed for drug
possession, Outlaw paid bail bondsmen to have them released from jail
as quickly as possible, usually within twenty-four hours, on a "Promise to
Appear" (PTA) in court. He met the bondsmen at a diner off the highway and gave them $5,000 to $10,000 to get his men sprung. Outlaw
estimated that over the course of his career, he paid a million dollars to
bondsmen for their services. But he hated them, calling them "cocksuck-
ers" and regarding them as profiteers from the same drug trade he was
involved in, but lacking the courage to expose themselves to risk. Often,
the scheduled court dates had a way of never actually occurring. Outlaw
would visit the witnesses and intimidate them to stop them from following through on the charges. One of Outlaw's regulars was Frankie, a white
addict who, when he showered and shaved, transformed into handsome
respectability. In the course of operations, the Jungle Boys had somehow
acquired a police uniform. They would corral Frankie, clean him up and
put him in the uniform, give him a gun and a holster, and tell him to "ap-
prehend" anyone whom Outlaw deemed a risk to his crew. Frankie was
skilled at impersonating an officer, and would somehow finagle to get the
targets into a car, at which point they would be held captive, under gunpoint, by members of the crews. Captives would be brought to Outlaw,
who, in an intimidating whisper that only he could pull off, would tell
them about the unspeakable things that would befall them if they carried
through on any threats to the gang.

There was only one final, but very important, instruction. "Don't sell
to Yale students," Outlaw told his troops from the earliest days of the
Jungle Boys. "It will bring too much heat."

Outlaw perfected this style of business in the gang's first two years, all before his sixteenth birthday. It powered the Jungle Boys into becoming the largest gang in New Haven, and made Outlaw a local celebrity wherever he went.

But Outlaw wasn't simply a strategist. On a daily basis, he operated as a kind of roving linebacker, moving freely throughout the Jungle at all hours. A lookout or bundle handler never knew when he would be watching. He often worked sixteen-hour days and barely slept. He intimidated people with his girth and size and then in the next moment hugged them and made them laugh. He was uproarious in a crowd, and then retreated into silence for hours, even days. You never knew which Outlaw you were going to get, and it kept everyone on their toes. He liked to tell people what to do. He found that most of the crew, fatherless like him, responded well to orders. It was almost as if he was a father figure, which was bizarre considering he was roughly the same age as most of the Jungle Boys. It was as if he was born for this, born to be a drug dealer, and born to be a leader of young men.

He made sure to look the part. He wore his hair close-cropped, never letting it grow into an Afro, which he considered outdated and soft. He found specialty stores for clothes big enough to fit him, and favored loud attire, the louder the better: designer sweats, orange suits, lime-green hats, maroon pants, crocodile shoes on a Saturday night. One day he drove to the Bronx, where in a storefront self-taught artists took molten gold and applied it to his incisors, inscribing a J on one tooth and a B on the other. It was ambiguous what the letters stood for: Juneboy or the Jungle Boys? No matter: they were becoming one and the same. Back in New Haven he smiled wherever he went, showing off his glittering new teeth for all to see. Dozens of gangsters, or wannabe gangsters, cast their teeth in gold after that.

In all of this, Outlaw had an ability to see things quickly, to read and react and respond. He seemed to have an uncanny sense of what might occur before it actually did. It was perhaps these same perceptual characteristics, beyond his physical abilities, that made him a special basketball player: the ability to see the whole court, and move into the right space at the right time. Outlaw's favorite player was Larry Bird. The Jungle Boys, who all loved Michael Jordan and Patrick Ewing, made fun of him for liking a gangly white player from Indiana who moved in a herky-jerky fashion. But to Outlaw, Bird was the greater player. Bird could anticipate patterns and lines of attack that were inaccessible to more physically gifted athletes.

But more than anything, it was the feeling of drug dealing that Outlaw loved. It was a feeling of mastery. Sure, it was magnificent to have huge wads of money in his pocket, and dozens of boys hanging on his every word, and countless girls available to him—Outlaw was still occasionally seeing Phaedra, but she was only one of many girls at this point—but what was much more satisfying was knowing that he was really good at something. *That* was what was so intoxicating. Because if he truly reckoned with himself, he had never excelled at anything before, unless you counted street fighting. He had been mediocre at school and been benched for his basketball team's state championship. Most everything he'd been involved in had led to hurt or punishment or pain. But now there was something electric and limitless and deeply potent about what he was doing. Conversation stopped whenever he walked into a room.

OF ALL THE new boys that surrounded Outlaw in the gang, three became his especially close and trusted associates. Each of these three boys, all of whom were a couple of years younger than Outlaw, eventually rose to a new level in the gang, that of lieutenant. This meant they were the seconds in command to Outlaw, and could make decisions if he was not available.

Pong was stout and built like a wrestler. He typically had a flat, neutral, almost soldierly look, until he broke up into laughter, which happened often. Pong had grown up in the Jungle, which conferred an immediate advantage to Outlaw, as Pong facilitated an invaluable entrée to the families that lived there. From an early age, Pong's ambition in life was to become a drug dealer, and he made his first sale when he was ten years old. Pong's father died young from drugs, and his mother was a hard-core, if secretive, cocaine user. To Pong, being a soldier for Outlaw was the realization of a dream. He thought of it as akin to making the NBA. Given the deficits of his childhood, he had never expected to be drafted into a gang as potent as the Jungle Boys. To Outlaw, Pong was the perfect infantryman, willing to do anything that Outlaw asked of him.

Alfalfa was small, almost petite, light-skinned. He was the youngest in the gang, thirteen when he first joined, and couldn't have weighed more than 120 pounds. Alfalfa had a philosophical and soft-spoken quality and vaguely resembled the movie director Spike Lee. As soon as he entered puberty, he sported a goatee on his sharp chin. Alfalfa and Outlaw grew up together at Q View, where Alfalfa's mother and Pearl had been very close. In fact, Alfalfa was related to Outlaw in some way that neither was entirely certain of. They believed that their grandmothers were sisters, but either way, they just decided to call each other cousin, although Outlaw often assumed the role of big brother. When Alfalfa fell into the gang, his mother came to Outlaw and said, "You're gonna take care of my boy, right?" "Yes, ma'am, I will," Outlaw said. Outlaw started Alfalfa as an unarmed lookout and wanted to keep him there. But Alfalfa turned out to be relentless in his ambition and skilled at money management and strategic planning. When the gang expanded to another location at a house in the Hill neighborhood, Alfalfa, along with Pong, was promoted to de facto manager of the new site.

For Alfalfa too, joining the gang was like entering a glittering world. He would show up after school at three and work until ten or eleven at night. He couldn't wait to get to work every day. A world had opened up to him, particularly a world with endless available women. Alfalfa was a father by age fifteen.

The third younger boy who joined the gang was Lion, with whom Outlaw was the closest, in what became a fraternal and loving bond. Lion's father too had been involved with drugs and crime, and had placed his son's hand on the trigger of a gun when he was ten years old so he could see what it felt like. Then his father disappeared from his life, leaving Lion's mother to raise the boy. Lion was tall, almost as tall as Outlaw, and powerfully athletic. He was from across town and went to a different school, where he was a point guard on the basketball team. Of all the members of the Jungle Boys, only Lion could match Outlaw's energy and power, and they became inseparable, to the degree that their relationship became a source of jealousy to other members. As the gang grew, Lion was exalted to a level of leadership and trust that no one else enjoyed.

When Alfalfa and Lion first joined the Jungle Boys they were still in high school. Outlaw felt guilty about getting them involved in the gang at such a young age. He would ask them to show him their report cards so he could be reassured that they were still taking school seriously. But after a while, with so much cash to be made, Lion and Alfalfa dropped out, about which Outlaw felt a twinge of guilt.

And finally there was Rodney, who had his own separate but related crew of twenty boys and men in the back of the Jungle. Rodney had his own suppliers and pricing, but he joined forces with Outlaw when it was expedient for both of them to do so, particularly when they wanted to make a show of force. Together Outlaw and Rodney had a combined crew of sixty members, an army that no other gang in town could remotely match. The distinctions between the two groups were lost on outsiders, and both factions were simply regarded as the Jungle Boys.

As the gang grew, Outlaw, like any good manager, invested in the business, from green camouflage clothes to weapons. He later estimated that he procured three hundred guns for the crew. Weapons were cheap, fifty or a hundred dollars each, and easily acquired from addicts who themselves had acquired them in house robberies. Black market gun dealers also came from New York to sell their wares. Outlaw bought Motorola handheld radios—police scanners—which gave the crew complete access to police communications over the public airwaves. The crew frequently knew beforehand when the police were going to patrol Church Street South. The Jungle Boys would put their stashes away and smile broadly at the officers. The gang acquired dozens of junky cars for a few hundred dollars each, which they parked in the back alleys of the Jungle and on side streets all over town. Outlaw and his lieutenants committed the cars' locations to memory, and if a Jungle Boy was on the run from the police or another dealer, they would jump into one of the cars, which were kept unlocked with a key hidden under the carpet, and tear away.

But most of all, the Jungle Boys had money. By 1991, the *New York Times* reported there were 5,000 to 7,000 intravenous drug users in New Haven, who along with other drug users were buying a million dollars' worth of product per day. Pong, Alfalfa, Lion, and other senior members often had $5,000 to $10,000 on their persons, and two or three luxury cars—tricked-out Cadillacs or BMWs or Mercedeses—to go along with their gold chains, and outrageous clothing, and the latest sneakers and hats. It was mesmerizing, fantastic, beyond a normal person's comprehension.

With typical generosity, but also strategic acumen, Outlaw shared the wealth with families living in the Jungle. If a family needed new furniture—and was willing to look the other way as the Jungle Boys plied their trade—Outlaw bought them dining room sets, leather couches, and the latest televisions. Delivery trucks would pull up within days and unload the goods. The crew shoveled sidewalks in the winter, raked the leaves in fall, and helped elderly residents with groceries. On summer nights, a festive party atmosphere prevailed over the Jungle. Alfalfa would set up boom boxes with massive speakers. Listening at maximum volume to N.W.A, LL Cool J, Ice-T, and Public Enemy, the crew would cook steaks and hamburgers on charcoal grills. Everyone was welcome. Girls might start to dance in front of a bonfire to the cheers of onlookers, and then older women would too. The children would throw a football or run races, and teenagers would wander off for trysts in the apartments or in cars. Older folks would sit in lawn chairs and drink and eat. On a hot night the party would go on until three in the morning. Good-natured fights might break out between some of the young toughs in the crew as the liquor and cocaine flowed, only to be laughingly broken up by Outlaw or another crew member. Newer gang members had only ever seen such things previously in B-level movies and rap videos, and now they were living in it. Everyone agreed it was crazy. Unbelievable.

It was in moments like these—the parties, the liquor, the humor, the camaraderie—that the entire crew found in one another the one big uproarious family that virtually all of them had lacked growing up. It seemed like the riotous fun would never end. Typically, Outlaw came up with the plans. "We're going to Harlem to go to the Apollo." Or "We're going to a rap concert in the New Haven Coliseum—we're buying sixty tickets. Bring anybody you want to!" As time went on, they went farther afield, to casinos in Atlantic City; to Virginia Beach, where they bought a house; and to the NBA All-Star game in Atlanta. In those wonderful moments, Outlaw would light a joint and preside over everything. All was right with the world. Looking at what he had created, he felt a surge of pride. He'd see a group of young boys laughing with delight at some inside joke, wearing clothes that no one in the history of their family could ever have been able to afford, and smile broadly. They had taken on the world, with everyone conspiring against them, and the battle and the spoils were theirs. Who cared about school or rules: they'd paid attention to none of it and they had won.

And the thing was: *no one had expected any of this.* When Outlaw started, his aspiration was maybe, if he was lucky, to rise to the level of another Butchy. He was the equivalent of forty Butchys now. Who would think twenty-dollar bags of white powder could fuel all of this— this madness, this money, this sexual potency?

They would die for each other if they had to. They would never sell each other out. In those moments Outlaw knew that the Jungle Boys would last forever.

But of course it was all built on violence and suffering.

When Outlaw looked into the mirror late at night after a sixteen-hour day, he knew, as much as he tried to delude himself, that all his achievements were the product of brute pain. He saw daily how the initial euphoria and overconfidence of cocaine turned quickly, often within fifteen minutes, into the drug's other side: depression, agitation, the not sleeping for days, the hallucinations, the permanent thousand-yard stare. Skeletal customers returned night after night, their pupils dilated, their noses bleeding, their arms or necks or feet scarred with track marks, their lips and fingers burned from smoking the drug. He would hear about customers who had "fallen out" with a heart attack. He saw men who would crawl across cut glass to get a fix, and women who would give a blow job to anyone in order to get high. Outlaw, in his zeal for profits and glamour, was largely indifferent to the suffering that surrounded him. Again, it was the price of business.

"The crack and cocaine epidemic uniquely devastated the community, particularly the black community," noted Stacy Spell, a policeman who walked the beat in Newhallville and who had tried in vain to bust Outlaw. "Before crack, there was a middle-class black community. Like my father, who worked in a factory in the fifties and sixties, and did pretty well. Houses were taken care of, the lawn mowed, the snow shoveled, a working car or two in the driveway. Working men would party on the weekend in neighborhood bars, but all they'd buy would be shots and beers. They were active in church and social clubs. But after the factories went to China and crack took over, it was like the bottom fell out of the paper bag. Half the houses were falling apart. Men would be strung out, having lost their jobs, and women were whoring on the streets." Indeed, in 1954, 33 percent of New Haven residents were employed in manufacturing; by 1977, it was 14 percent. The crime rate rose sixfold over the course of the 1960s, and another fourfold during the 1970s. The murder rate exploded from twelve in 1985 to thirty-four in 1989. In 1994, *GQ* published an essay called "The Last Boola-Boola" ("Boola Boola" being Yale's fight song), which described New Haven as a "war zone of poverty, crime, and drugs as frightening as any American city."

When it served his purpose, Outlaw was entirely comfortable contributing to that violence. He thought nothing of punching a crew member's jaw if he disrespected him. If one of the Jungle Boys talked in a movie theater, he'd cuff him and laugh hysterically. If any opposing gang member talked trash about him, Outlaw would personally demolish them with his fists, leaving them all but unconscious on the sidewalk. Or he would instruct one of his boys to do the beatings. Either way, the message would be delivered.

WITH HIS NOW overwhelming numbers and his role as an enforcer established, Outlaw began to do so much business with the Albanians that he purchased up to twenty-five kilos a month, at a cost of more than $400,000.

The process worked as follows: Weekly or biweekly, Outlaw would go to random pay phones and page the Albanians. The Albanians knew the call was from Outlaw because of the 203 number—Connecticut's area code—and typically would call back within minutes. In rapid-fire conversation, the Albanians would name a time and address in some obscure hinterland of Queens or the Bronx. It was always a different address. The next day, or soon thereafter, Outlaw—or as time went on, Alfalfa or Pong—would meet the Albanians in a clandestine apartment in New York and hand over the agreed-upon figure—$100,000, sometimes $200,000—in thousand-dollar bundles. The Jungle Boys would return to New Haven with only a verbal promise that the Albanians would make good on the deposit.

One might think it improvident to deliver all that cash on only a verbal pledge, but without exception the Albanians would appear a day or two later in New Haven, at the address and time that Outlaw had specified. To avoid risk, the drops were never made at the Jungle but rather at one of the increasing number of tattered apartments the Jungle Boys rented or otherwise had access to around town. The Albanians would pull up in unremarkable-looking cars—their only distinguishing feature being New York license plates—and walk briskly up the stairs, meet Outlaw and his security guys, and hand over the kilos.

There was never any need for security. Everything on both sides transpired in a cordial, respectful, even warm fashion. The fact was that Outlaw liked the Albanians with their strange accents. They were cool and slick and exotic to him, and he, perhaps, to them. The transactions proceeded in a manner that was nothing like what is typically portrayed in movies or on television, where every moment is saturated with tension, and teetering on violence.

But trouble did occur on those rare occasions when Outlaw would call New York and the Albanians would apologize, the sincerity apparent in their voices, saying that they were out of cocaine. "But please call back next week, maybe things will be different," they would say.

Such situations only happened half a dozen times, but they forced Outlaw into uncomfortable freelancing territory. After that one barren period, Outlaw vowed that the Jungle Boys would never be out of product again, and therefore, reluctantly, with a pang in his stomach, he would bring his most coolheaded associates and drive to New York City, to Harlem or Washington Heights. The moment the Jungle Boys got out of their cars with their gold ropes, sunglasses, and hip-hop gear, the calls came cascading down from young men sitting on the stoops of brownstones or standing outside bodegas. If they were in Washington Heights, it would be "Hey Papi, we got something for you. Come right this way, *hombre*." In Harlem, they heard, "Yo, man—you need something? You need some good shit?" Outlaw, taking care not to make direct eye contact, would nod and the crew would be ushered down a side street and whisked to a crumbling brownstone. They would walk up the steps into a narrow tiled hallway smelling of urine and huddle into the elevator.

The Jungle Boys would go to an upper-floor apartment that was fully dedicated to the drug business, in effect a minor factory. The shades would be pulled down, and strung-out-looking men and women with hospital masks would be breaking down kilos. Young guys with deadened eyes looked on, clutching AK-47s. The apartments were stifling hot in the summer and freezing in the winter.

Outlaw took the lead. "Looking for six kilos. We got cash."

He always instructed his crew beforehand what their roles would be—who would handle the money, who would talk, who would not talk, who would stay in the car. One ill-advised word, even a wrong look, could get them killed. Outlaw could just see the headline in the *New York Post* the next day: "Nine Men Found Dead in Apartment."

After one such transaction, in which everything had proceeded with perfect civility, Outlaw and his crew, with their newly acquired kilos, walked to their cars a block away. Just as they got into the seats, a group of four young men approached them.

"Yo, man, what you got there? You gonna give us some?" they said leeringly.

One of them pulled out a gun from underneath his athletic jacket and pointed toward the Jungle Boys' heads. The Jungle Boys slammed the gas pedals of their cars and tore away down the block, stray bullets flying past them.

But in the midst of all of this violence, something beautiful happened. Outlaw had continued to see Phaedra on and off, and she became pregnant. Outlaw was too busy to see much of Phaedra during the pregnancy, but he was there when she gave birth. He held the baby, a beautiful girl, in his hands. Phaedra beamed. Even Pearl was happy. Outlaw was a father at sixteen and the circumstances were far from ideal, but the girl was gorgeous and a gift from God and Pearl was a grandmother. Everyone agreed that the baby would live with Phaedra and Phaedra's mother, and that she would be named Marquetta.

While Phaedra and Outlaw both doted on the baby, and he provided plenty of money for the best baby clothes and toys over the first years of Marquetta's life, the relationship between the two parents steadily worsened, becoming irrevocably damaged after Phaedra's brother stole $27,000 from Outlaw. Outlaw had placed the cash in a safe that he kept at Phaedra's house. He visited the baby one day and found all the cash gone. When Outlaw asked Phaedra about it, she equivocated and then disappeared for a few days. The next time Outlaw saw Phaedra's brother, he was wearing gold ropes and clothes that he would never have been able to afford on his own. Outlaw smacked him in the head with a gun, and threatened to go after Phaedra too.

When the police jumped him in the Jungle and charged him with threatening, breach of the peace, and criminal mischief, Outlaw was stunned. He couldn't believe that Phaedra had actually filed a report on him. He was booked at the police station where his mug shot was taken, didn't bother to ask for defense counsel, and was sentenced to a three-week stay at the Manson Youth Institution in nearby Cheshire. Pearl was happy and relieved: maybe, just maybe, this time her son might learn something.

He arrived at Manson on May 11, 1987. He had just turned nineteen. The intake report from that day says: "he doesn't use drugs. Tried pot, but doesn't like it. Drinks wine coolers ([for] about 2 years, since they first came out). Drinks about 15 on the weekend . . . he says he does not have any particular plans for a career. His mom wants him to work at Winchester and [he says] 'maybe I will, or do machinist work.'" The three-week sentence was extended after Outlaw asked one of his girlfriends to bring him pot and was caught by the guards. The sentence was extended further when Outlaw racked up dozens of disciplinary tickets. The following is typical: "Inmate Outlaw gave this officer the middle finger from the dayroom of J cottage. . . . I asked subject what the insult was for and Outlaw stated, 'next time we will get physical.'" The three-week stay turned into four months, and he was released and picked up by Pearl in September. On the drive home to New Haven, she asked if he had learned any lessons. "Of course, Ma," he said.

He bounded down to the Jungle the next morning only to find the crew standing around. None of them was plying the trade that Outlaw believed he had so carefully taught them. But on the other side of the Jungle—Rodney's area—business was fervent.

"Why the fuck aren't you selling? Rodney is selling!" Outlaw shouted. The crew mumbled various excuses.

"You mean I've been fucking locked up for your asses and you haven't done nothing? What did I go to jail for?"

Outlaw knew from Suzie Jones's example how precarious the drug business was.

"All right, this is what I'm going to tell y'all. We starting up tomorrow morning at eleven, just like we used to," Outlaw said.

Meanwhile a few older men were selling in the Jungle Boys' spot. Outlaw walked up to them. "Just so you know," he said. "We starting up tomorrow morning. Don't you dare show up, or I'll take care of you. That's not a threat, that's a promise."

Rodney lent Outlaw two kilos, and Outlaw and Ricky and others spent the night cutting it with razor blades. At eleven the next morning, Outlaw was organizing the crew for action, when he saw one of the older men still selling on the sidewalk.

"I thought I told you not to be here."

"It's a free country."

"No, it's not."

Outlaw, almost twice the man's size, slammed him on the head with his fist. The blow staggered the man, buckling his knees, and he fell to the sidewalk, dazed and shattered. Picking up the twenty-dollar bills—the detritus of his drug dealing—that had scattered on the sidewalk after he'd been hit, the man unsteadily walked away, never to return.

After starting up again in this fashion, in what would be the final year of the Jungle Boys, Outlaw's empire expanded multiple times. He conducted a side business selling cocaine near his first location on a street corner in Newhallville. Half a dozen Jungle Boys were designated to work a spot on the corner of Winchester Avenue, and on a good day it could net $5,000 or $10,000.

Then a third location, a house on the corner of Baldwin Street and Congress Avenue in the Hill, fell into the Jungle Boys' lap. A small independent crew had sold cocaine from the house. The leader was an older man named Darryl, who sometimes bought cocaine wholesale from the Jungle Boys. At one point Darryl fell into a sizable debt with the Jungle Boys, in the range of $100,000 that he was unable to repay. It was unthinkable that Outlaw would let a debt of that size stand, and he went to look for Darryl only to learn that he had fled town. In a form of settlement to avoid violent retaliation, Darryl's remaining crew chose simply to let the Jungle Boys take over their house. Pong and Alfalfa were selected to manage the new location, which was on a major four-lane commercial artery and had a side window through which customers could be served, McDonald's-style. Baldwin Street turned out to be an unexpected cash cow, even rivaling the Jungle in profits. Some days Baldwin might make $30K, and the Jungle $40K. Outlaw used the competition to his advantage. At the end of the day he'd take the leaders of both establishments to a diner and tell them who was ahead. "See if you can beat the Jungle tomorrow," he'd say to Alfalfa or Pong.

Outlaw found something deeply satisfying about having multiple operations. There was so much money flowing now that the gang didn't know what to do with it. A room in the Jungle where they stored cash was now overflowing with bills. The Jungle Boys took to burying their money in the ground in isolated spots in New Haven. In the coming years, when Outlaw paid defense attorneys more than $100,000 cash, the bills had dirt on them.

THERE WAS ONE remaining crowning glory to be accomplished in Outlaw's gang career, and that involved taking the fight to the police and beating them at their own game.

One night, Outlaw was at a bank of pay phones in the train station, calling the Albanians. An officer tapped him on the shoulder, saying, "I see you're making a lot of money in the Jungle."

Outlaw said nothing.

"You're making *a lot* of money," the officer continued. "I want you to meet me here every Sunday night and give me five thousand dollars. You give me five grand a week, and I won't give you any problems."

Disgusted, Outlaw strode toward the exit doors, which swung open automatically. The officer followed Outlaw onto the street. The two stood across from each other in the black of night, the Jungle across the way. Outlaw had just acquired a .41 Magnum, a gun whose beauty and power he had fallen in love with. It was silver with a wooden handle, eleven inches long with an outsize barrel. He pulled the Magnum out of his pocket and pointed it at the officer's head.

"You see this? If you want your money, come and get it!"

The officer slunk away into the darkness.

A week later, Outlaw was being driven around town by one of the crew when he spotted the same officer in plain clothes with his wife and kids in a white sedan. Outlaw told his driver to pull up next to the sedan. When the two cars were side by side at a red light, Outlaw rolled down the window and pulled out the gun.

"You still want that money? Well, here it is. I'll blow all your fucking heads off."

The family—the wife, the young kids—looked aghast, their mouths open. The officer drove away furiously as soon as the light turned green.

Not long afterward, one of the few African-American police officers on the force came to visit Outlaw. The cop said, "I understand you had a little run-in with one of our officers. He's well respected and I'm here to tell you that we don't want any problem between you guys."

"I don't have no problem with you," Outlaw replied, "but I got a problem with him. He's dirty as fuck!"

Ignoring Outlaw's accusation, the cop simply reiterated that the police didn't want any problems.

On another occasion, a youngish officer who apparently didn't know any better made the mistake of wandering into the Jungle alone. Seeing Alfalfa handling some bags of cocaine, the policeman ran over and quickly overpowered him. After cuffing him, the cop began reading Alfalfa his Miranda rights.

Outlaw spotted all of this from across the yard and charged over, accompanied by five of the Jungle Boys.

"What the fuck you doing to Alfalfa?" Outlaw said to the astonished policeman. "What the fuck you arresting him for? What did he do wrong?" The policeman mumbled something about arresting Alfalfa on suspicion of cocaine possession.

Outlaw cut him off. "Get your motherfucker ass out of here," Outlaw said, pulling out his Magnum. The other Jungle Boys, as always following Outlaw's lead, pulled out their weapons and the policeman soon found himself in the center of a circle with six guns at his head. The cop ran off to the safety of the street without even releasing Alfalfa from his restraints. Thankfully, the Jungle Boys had plenty of handcuff keys lying around.

BUT OUTLAW WASN'T always hostile toward the police. He was only too happy to help out officers if they were like him: "straight up."

One night in October 1986, Stacy Spell, the Newhallville beat cop, was walking down a street in his district. Donald Fagan, a member of a rival gang of the Jungle Boys, pulled out a shotgun and fired twice at Spell from twenty feet away, a sure kill. But somehow his gun didn't discharge. Spell shot back and critically wounded Fagan, who took months to recover. Later that night, when the ballistics team at the police station tested Fagan's gun, the weapon discharged normally. Due to a mechanical error, or perhaps the grace of God, Spell had escaped certain death.

In the aftermath of the incident, Spell felt traumatized and also oddly guilty: grateful but also confused as to how—and why—he had survived. Suddenly he was beset by flashbacks, filled with doubt and anxiety, and found it impossible to do his job. He took a leave of absence from the force for a few months and went to California to visit his children, who were with his ex-wife. When he came back to New Haven, he tried to get back into a routine.

Outlaw had always liked Spell, regarding him as his brother on the other side of the law. They were similar in a lot of ways. They resembled each other in their size and physical power and grew up in the same neighborhoods and in many of the same circumstances. Spell's uncle was a major heroin dealer and Spell had lost six relatives to the streets, either to drugs or violence. Spell believed he easily could have gone the same way as Outlaw, had he not entered the military right after high school. Spell too admired Juneboy from afar. "Business was business," Spell says. "They did their thing, and I did mine. But I always respected Juneboy. He was impressive, with a great sense of humor. Under different circumstances we could have been friends."

At one point during his leave, three months after Fagan's shooting, Spell ran into Outlaw at a movie theater. Outlaw was surrounded by an entourage of ten Jungle Boys, who were hanging on to Outlaw's every word as usual.

"Outlaw came up to me, gave me a huge hug," Spell recalls. "Of course, the other guys followed his lead, and so they embraced me too. 'Hey big Stace!' they said. 'You want us to get him? You want us to get that fucking guy Fagan? We'll take care of him for you!'

"I couldn't believe it. Here are these gangsters looking out for me, a cop! Of course, I said, 'No, no, I'm fine. I don't need anything.' But then I went back to my car and started bawling. I just couldn't stop crying."

What Spell didn't know was that a hit had been called on him by

Donald Fagan's gang in retaliation for the shooting. Outlaw had embraced Spell because he liked him, and also to send a message to the opposing gang: mess with Spell, and you mess with the Jungle Boys. Subsequently, the hit had been cancelled. In a certain way, Outlaw had become more powerful than the police.

IN THE CREATION of the Jungle Boys, Outlaw's timing was exquisite, as it would be at many other times in his life.

If one was interested in forming a gang in New Haven at any point in the twentieth century, there was probably no better time to do so than the mid-1980s, when there was a near-void in organized criminal activity just waiting to be filled. From the 1950s through the 1970s, the crime scene in town was dominated by the Mafia and in particular by two men who represented opposing New York families that had outposts in New Haven. Ralph Tropiano represented the Colombo crime family, and Salvatore Annunziato the Genovese family. They despised each other and were opposites, physically and in terms of personal style. Annunziato—a stocky, brash, and tempestuous former boxer with a drinking problem who clocked in at barely five feet tall—ran the unions, while Tropiano—tall, quiet, bespectacled, almost professorial—dominated the bookies and the numbers game. Annunziato was convicted by the Feds in the bribing of a highway contractor. After his release from prison in 1979, he disappeared, presumably murdered, though the case has never been solved. Tropiano was convicted of bribing police officers to protect his gambling operation. In prison he ratted out twenty-one Mafia members to the FBI, and after his release was shot dead on a Brooklyn Street in 1980. With Tropiano and Annunziato gone, the mob scene in New Haven dwindled to a sliver of its former self, amounting to small-time numbers running, fencing stolen goods, and loan-sharking. Similarly, while there were plenty of small operators—stickup men doing robberies or lone-wolf pot and heroin dealers—no substantial African-American criminal groups emerged in the sixties and seventies. Outlaw arrived to a relatively clean slate.

And just as he arrived, an explosive element came along that changed the New Haven—and American—drug scene forever. Cocaine had enjoyed an early popularity at the turn of the nineteenth century among a small segment of the population, including members of the intelligentsia, and was briefly a legal drug, famously used as an ingredient in Coca-Cola. But Congress passed the Harrison Narcotics Tax Act of 1914, making the drug illegal and putting an end to its at times widespread use. As recently as the late 1970s, many experts and public health officials believed that cocaine was a relatively benign substance. "To be a cocaine user in 1979 was to be rich, trendy and fashionable," says Mark Kleiman, a UCLA professor of public policy. "People weren't worried about cocaine. It didn't seem to be a real problem."

But in the early 1980s, two underlying factors in the cocaine business shifted. The means of production in South America, where the coca plant was grown, was transformed from a cottage industry of small groups of farmers into a major business bankrolled by organized families, or cartels. With that change, the profits—now billions of dollars, not millions—became substantial enough that governments and legal systems could be bought off. The borders opened up and cocaine flooded into the United States, by far the world's biggest market. At the same time, the drug itself became much more potent. Snorting the drug was the primary way to use cocaine through the 1970s, but nasal administration is relatively inefficient, with the drug having a comparatively slow arrival time to the brain. By the 1980s, users began to inject cocaine intravenously and inhale the drug in new smokable forms, crack or freebase, which produced more potent and immediate highs. Crack was made by dissolving the powder in water, adding baking soda, and boiling and filtering the combined ingredients to create small nuggets. Freebase cocaine involved the use of ammonia to liberate the drug from the salt form in which cocaine is naturally found, creating an almost 100 percent pure version of the drug that hits the bloodstream within seconds.

Driven by these structural changes, suddenly cocaine was everywhere, and portrayed eminently as the drug of its time, glamorized in rock and rap songs, television and movies. The drug even looked sexy—a snow-white powder often shown being snorted through rolled-up hundred- or thousand-dollar bills by beautiful men and women. While the purpose of marijuana and LSD, the prevailing drugs of the 1960s and 1970s, had largely been to escape reality or create a new one, cocaine was viewed as a performance enhancer. It was the perfect drug for the high-flying 1980s, an ideal complement to Reagan-era testosterone, and for a brief moment, cocaine was the first illicit drug in decades not to be broadly stigmatized. By the mid-eighties, as many as eight million Americans used cocaine regularly, and between 1985 and 1988, the number of people using crack or cocaine at least once a week rose by a third. The American cocaine market by the end of the decade was valued at $140 billion, over 2 percent of the gross domestic product, and larger than many legitimate sectors of the economy. "The average drug trafficking organization, meaning from Medellín to the streets of New York, could afford to lose ninety percent of its profit and still be profitable," said Robert Stutman, a DEA agent.

But the glamour narrative that surrounded cocaine collapsed with the arrival of crack. Most of America didn't know what crack was in the early 1980s; by 1986, driven largely by the hysteria of white politicians, it had become a household word. Crack became tied overwhelmingly in the public's mind to black people, ghettoes, thugs, and the emerging genre of gangsta rap. The untamed menace of crack was brought to new heights after the death of basketball star Len Bias in 1986, two days after he was picked second in the NBA draft. Bias overdosed on a mixture of alcohol and cocaine, but likely because he was black and young, the rumor spread that crack had killed him. A few weeks after his death, both Republicans and Democrats forged the Anti–Drug Abuse Act, which fundamentally transformed the law enforcement community's response to drug abuse from a rehabilitative approach to a punitive one. Soon afterward came further legislation, which called for new prisons and mandatory minimum sentences for drug violations. The new laws made punishments related to crack up to a hundredfold higher than for cocaine. Nancy Reagan, who led an anti-drug campaign from the White House, deemed the legislation a personal victory.

In an odd way, it was a personal victory for Outlaw too. On the streets of New Haven, Outlaw found that the more sensational, negative press crack and cocaine got, and the more laws were passed and criminal penalties went up, the more people seemed to want the stuff. As cocaine, crack, and freebase dominated the airwaves, Outlaw's profits soared, and his customer base grew and diversified to include more whites, suburbanites, and professionals of all colors. After Len Bias and Nancy Reagan, it was not at all unusual to see a man in a suit and tie in the line, or a woman waiting for a fix on her way to work.

Outlaw was completely unaware that he had created something unique, something that had not been done before at a national level. According to *African American Organized Crime: A Social History*, by Rufus Schatzberg and Robert J. Kelly, black street gangs of the crack era were largely characterized by three phenomena: widespread recreational use of drugs by members; engagement in high levels of petty crime; and residing in and operating out of a single territory while defending that territory as necessary. Schatzberg and Kelly wrote that the Jungle Boys broke this paradigm on all three fronts, making them more successful than their counterparts and infinitely harder for law enforcement to catch. Other than Rodney, the Jungle Boys did not live in Church Street South, and some even lived outside New Haven. The crew concentrated on the big prizes of drug dealing and the enforcement associated with it and engaged minimally in other crimes. It was highly unusual too for a gang to be as "civic-minded" as the Jungle Boys were: to pay for food and parties, July Fourth fireworks, and to keep the housing project immaculately clean. Without even knowing it, Outlaw had created a national model.

JUST A FEW years later, when he was in prison, Outlaw would think of the following episode as the highlight of the Jungle Boys. On a raw March morning sometime toward the end of the Jungle Boys' run, the sky a battleship gray, Outlaw was patrolling the perimeter of the Jungle, readying for another day of dealing in the howling New England winter. One of the Jungle Boys, Ray Ray, pulled up in a car.

"What up, Ray Ray?" Outlaw said.

"Going to Florida."

"What?"

"Going to Florida."

"Give me a minute," Outlaw said.

Outlaw ran into an apartment, packed a suitcase, hopped in the car with Ray Ray, and drove to LaGuardia Airport. Three hours later, after drinks and a movie on the plane, they touched down in Orlando. They rented a huge boat of a car and checked into a pink stucco hotel on International Drive, populated by wandering flamingoes and close to the Disney theme parks. Outlaw and Ray Ray stayed for a week, drinking Dewar's and Bacardi and swimming in the aquamarine-colored pool. Outlaw paddled around for hours. Tourists stared at them wherever they went. Who was the big dude? A lineman for the Dolphins? A rapper? A boxer?

"Let's stay awhile," Ray Ray said. "Let's make it a little vacation. We've been working for a long time."

Outlaw nodded in agreement. *I've been working four years straight. I haven't stopped for a day.* They checked out of the hotel and rented a condo for a few months. Outlaw ran the Jungle Boys remotely, or let Pong and Alfalfa oversee things, while Ray Ray partied with call girls. Outlaw wasn't interested in paying for sex, and became friends with a girl he met at the 7-Eleven.

Outlaw found Florida to be a psychedelic experience. The tropical hues felt like a buzzy dream. He spent days driving around the state. He loved the highways, which were flat and smooth like elevated rivers, and lacked deep potholes from the frost heaves of winter. As he took in the unfolding landscape—the sugarcane and orange plantations, the alligator swamps—he felt the strain of the last four years dissolving from his body. He felt proud of his success. He could never have experienced anything like this had he finished Wilbur Cross. He thought of the idiots who had stayed in school, buying into a clearly failing system, now working in dying factories or Cumberland Farm convenience stores, or unloading trucks. Here he was looking at palm trees and driving past ten-million-dollar Florida mansions and sleeping with whomever he wanted, a plastic bag holding $40,000 by his feet on the car's floor.

One very early morning Outlaw found himself driving toward Tampa, when he remembered that the New York Yankees held spring training there. At a gas station, he asked for directions to the stadium. When he pulled into the complex, it was still only eight o'clock and no one was around. Outlaw got out of the car, looked around, and noted the beauty of the baseball stadium, a miniature version of Yankee Stadium. A luxury bus pulled up, and out streamed the Yankees players. Many of them were big men but none was as big as Outlaw. The players began congregating around him, sizing him up as a fellow athlete or celebrity.

"How ya doing?" the ballplayers said. They all exchanged high fives and fist bumps. Outlaw never said who he was, nor did they ask. They just knew he was somebody important.

Years later in prison, he would think about this scene and smile through his tears. He had indeed become that guy whom everyone wanted to be, just like he'd pictured it when he was at Long Lane. It was like an hallucinatory scene in a movie in which he had starred.

CHAPTER 3

Kill the King

Outlaw in a custom-designed Dapper Dan gold suit, Virginia Beach, 1988.

As HIS EMPIRE GREW AND GREW, A PHRASE BEGAN TO ENTER OUTLAW's head: "King of kings, boss of bosses." He didn't know where it came from exactly, but he liked that phrase. He liked the rhythm of the words and the way they sounded together.

Everyone talked about him as a drug kingpin or major "gangster," but Outlaw disliked the words *gangster* and *gang*, preferring the term *organized crime*. The word *gang*, he believed, didn't adequately capture the forethought and planning that went into his creation. In New Haven and elsewhere, *organized crime* was a term reserved almost exclusively for the Mafia, which Outlaw thought was unfair, if not racist. It implied that black kids couldn't be organized and he was proof that wasn't true.

But at the same time, another phrase began to go through his head: *Kill the king.*

"Kill the king, and you're the king": that was one of the tenets of the streets. Outlaw sensed that he was going to be a victim of his success, that it was only a matter of time before he was going to get shot. He increasingly found himself tuning out of conversations to scan the street, or a house, or a car going by, looking for snipers or people coming out of the shadows with revolvers and machine guns. He even began to pray that when he did get shot the bullets would take him out entirely and not maim him for life.

Concerned about retaliation, he decided he needed a private place to stay, and found a sublet of a sublet through a friend of a friend. It was a beautiful apartment, five blocks from Yale. On any given night, only Ricky would know that he was there. In the building were graduate students and professional people, and families with babies in strollers, and children in school uniforms. It was as if Outlaw had been suddenly displaced into another world, an entirely distinct and secure part of the city. He toned down his lifestyle in the apartment, kept more conventional hours, wore more muted clothes, and brought a girlfriend there only once or twice. The tenants sensed that he didn't exactly belong, but with his charm and energy, he greeted them warmly in the hallways and won them over. The place was tastefully decorated with a glass coffee table, oak dining table, and leather couches. Outlaw became a quiet homebody there, enjoying his time away from the gang.

As if to craft a further veneer of respectability, he started a new relationship with a more conventional kind of girl. He was at a diner with Ricky when he spotted Camile Leslie, whom he knew casually from high school, four booths away. She'd grown up near him in Newhallville. Outlaw asked Ricky to ask her to join them. She demurred but gave Ricky her number. Later, Outlaw invited her out on a date. Camile was in college now, a nursing student in North Carolina. She was entirely different from the girls who partied around the Jungle. Camile was quiet and self-effacing, and kind perhaps to a point of weakness. She was a churchgoing and dutiful girl from a strict family who maintained nightly curfews. For her part, Camile was scared of, but also attracted by, Outlaw's flamboyance. She was drawn by the excitement that always seemed to surround him, which she remembered from their first year at Wilbur Cross, even though they had barely talked. Camile and Outlaw had a good time. They went out to the movies and dinner, always to the best restaurants, drinking the best liquor, and driving the best cars. Outlaw seemed almost relieved to be away from the insanity of the Jungle. He opened the car door for her, always paid, brought her roses, and took her to Harlem on her birthday. Camile began to think of Outlaw as a nice guy, just a big teddy bear. She knew exactly where all the money came from, but she chose to ignore it. Camile also knew there were other women. Rumor had it that Outlaw in fact had fathered three more children with other women after the first baby with Phaedra, and that even more kids were on the way. Even when Camile found those rumors to be true, she chose to ignore them.

THE LAST SUMMER of the Jungle Boys was the summer of 1988, and it began auspiciously enough. During his trips to New York, Outlaw found himself increasingly drawn to a bodega on East 125th Street. Actually it was a former bodega, transformed into what was called Dapper Dan's All-Night Clothing Store. The place never closed, staying open 24 hours a day over eight years, and it was run by Dapper Dan himself, a forty-ish-year-old fashion impresario who just about single-handedly invented hip-hop fashion by reimaging and repurposing Gucci and Louis Vuitton knockoff fabrics, which he sold to Arsenio Hall, Mike Tyson, LL Cool J, Run-DMC, Salt-N-Pepa, and Bobby Brown. Outlaw frequented Dapper Dan's often, and the two men, sharing an outsize sense of style and an entrepreneurial flair, became close. Dapper Dan, or "Dap," as Outlaw called him, was as slight and thin-boned as Outlaw was big and bearlike. Outlaw spent tens of thousands of dollars on Dapper Dan's clothes. For his part, Dap was experimenting with gangsta fashion, and to have an actual gangster like Outlaw as an associate added to the store's ambience and street cred. One night in the early summer of 1988, Outlaw decided to outfit the members of his inner circle with custom-made gold Dapper Dan suits and pay Dap $14,000 for it. It was a big enough request that Dap personally drove down to Virginia Beach, where the Jungle Boys had a house, to drop off the suits and take pictures of the occasion. To Outlaw, the night was magical. In his gold lamé suit, he looked and felt like a king as he looked out over the beach and took in the breezes rolling in off the Atlantic. Sometime later, on June 27, Outlaw and his crew attended the Mike Tyson–Michael Spinks heavyweight championship fight at Trump Casino in Atlantic City, where they sat in the fifth row among movie stars and other celebrities. A few seats away was an NFL player, an All-Star and a household name. Before the fight, the Jungle Boys had given the football star free cocaine. Tyson demolished Spinks ninety-one seconds into the first round. In the commotion after the fight, the football player desperately followed the Jungle Boys out of the convention center, plaintively calling after them, "Do you got any more coke for me? Do you? Do you?" The Jungle Boys thought this was hilarious, and partied late into the night on the boardwalk. Stoned, they drove back to New Haven in Rolls-Royces they had rented just for the occasion. They were on top of the world.

But then a series of violent episodes occurred. The first was what Outlaw came to think of as the July war.

Once, during one of Outlaw's visits to Shelton Avenue, Pearl said that some boys at Q View were threatening Outlaw's brother. A gang had since formed at Q View, calling themselves the Island Brothers, or simply the Island. The Island was run by an up-and-coming gangster known as Blue Eyes. Outlaw knew him well. They had grown up together at Q View. Blue Eyes' family life had been unstable, and good-hearted Pearl had taken him under her wing from time to time. Rodney's mother too had taken care of Blue Eyes.

At the kitchen table, Pearl said she didn't really know the details, but someone in the Island had been threatening Outlaw's older brother. Could Juneboy do something about it?

Of course he would. Outlaw would do anything for Pearl and for his family. He promptly went to Q View and talked to the men at the picnic tables in the courtyard and asked if they knew who was threatening his brother. The men named two of Blue Eyes' guys.

A few days later, Outlaw and Ricky saw these two men getting into a parked van on the corner of the Green, outside a new mall built in an attempt to attract people downtown. A Macy's department store had recently opened. Hundreds of people were passing by on the sidewalk but this did not deter Ricky and Outlaw from pulling up beside the van in which the Island Brothers sat and blocking their exit. With his long-legged athleticism Ricky bounded into the van and smashed the driver on the side of his head with the butt of a pistol.

"Yo, don't fuck with Juneboy's brother," Ricky said.

"Take that as a final warning," Outlaw said.

He and Ricky got back in their double-parked car, which was snarling traffic. Nothing happened for a few weeks afterward and Outlaw thought the business was over. But then a few weeks later one of Outlaw's crew came back to the Jungle, saying that he'd been selling cocaine near the projects where the Island Brothers were based—as a freelancer, he admitted—and members of the Island said they were going to kill him.

Okay, that's it. We gonna take care of this, Outlaw thought.

He instructed the crew around the Jungle, twenty or so guys, to suit up. They put on bulletproof vests and camouflage pants and shirts to convey this was an actual war and had the dimensions of a military operation. Outlaw put on a black woolen ski mask.

"Why the fuck you doing that, Juneboy?" someone said. "It's fucking July. It's hot outside." The guys all laughed at him.

"The joke's gonna be on you guys. They gonna say in a court of law, did you see Juneboy's face? Can you identify him? They won't be able to," Outlaw said.

"But everybody can tell who you are. Don't matter if you're wearing a mask or not."

"Like I said," Outlaw responded. "It won't hold up in no court of law."

They piled into their cars, four or five men in a vehicle. Some of the crew took motorcycles. Before they left, Outlaw gave them the plan.

The main thing—like all of Outlaw's big moves—was it had to be quick. Outlaw said they had to get back to the Jungle afterward, zipping down the interstate at a hundred miles per hour, and make like the whole thing never happened.

They drove the mile and a half to Q View and parked in the back. Outlaw told the crew to climb to the roof of the one-story social hall, which was easily scalable by hopping on top of the utilities shed in the back. They all clambered up, guns strapped to their backs. Below them was the courtyard between the barracks-like apartments. Members of the Island Brothers sat at picnic tables, smoking, drinking beer, completely oblivious. It was late afternoon, broad daylight.

The Jungle Boys fired a warning shot to allow anyone who didn't want to be there to escape. And then, on top of the social hall, the Jungle Boys began firing in earnest. The bullets sprayed the dirt and ripped apart the picnic tables. All the Island Brothers successfully fled into alleyways and apartment corridors. Remarkably no one was hurt.

The crew got back into their cars and on their bikes and sped back to the Jungle. Before even seven minutes passed, the entire gang was out on the hill by the pine trees. Outlaw checked on the police scanner, through an earpiece: "There's no way it could have been Juneboy, Sarge," the dispatcher said. "I see all of them on the hill. They couldn't have gotten back that quickly."

The Jungle Boys spent the next hour or so celebrating their decisive move, drinking and carrying on with booze and joints. Junior members of the gang set up grills and cooked burgers and dogs. A triumphant atmosphere descended on the Jungle that night: they had wanted to send a message to the town, and they had.

Outlaw was smoking reefer and generally feeling good, when one of the lookouts charged up to him. "Yo, Juneboy, your mother is here!"

"What the fuck?" This was the most shocking development yet. Indeed, Outlaw saw Pearl on the sidewalk below. She was with Rodney's mother, and, incredibly, they were accompanied by a fearful-looking Blue Eyes. Outlaw took one look at his mother and sprinted away. He was perfectly capable of facing down Blue Eyes and his crew, but there was no way he could handle Pearl, the matriarch. Taking on men with guns was one thing, but confronting Pearl was another.

Outlaw went to his Yale-area apartment and returned to the Jungle the next morning. "What the hell happened last night?" he said.

Alfalfa and Rodney explained that Blue Eyes had pleaded for a truce. He brought the mothers along for help.

"And what did you guys say?"

"Out of respect for Miss Pearl, we said we'd lay off for a while."

Outlaw agreed. Everyone needed to comply with what Miss Pearl said. But he and the whole crew had diminished respect for Blue Eyes. His plea was widely considered a "mama's boy" move. Blue Eyes may have been a big-time drug dealer, but he was vilified unmercifully after that.

During the rest of that summer, whenever Outlaw was in his mother's home—catching a nap or unable to resist her home cooking—Pearl would pester him. "Juneboy, why'd you go out to Q View to go after the Island Brothers? That's where you grew up. That's where I raised you."

"I don't know nothing about that, Ma."

"Don't lie to me, Juneboy!"

"How'd you know I was out there, Ma? No one ever identified me."

"You were there, Juneboy, I know you were!"

"Don't worry about it, Ma. No one saw my face that day."

Which was technically true because he wore the black mask during the entire shoot-out.

Sometime later, Outlaw was informed by a rival in Newhallville that he and the Jungle Boys were getting "too big for their britches" and were being banned from the Oasis nightclub, the most popular black club in New Haven. People came from all over the state to party at Oasis.

To Outlaw, clearly another show of force was necessary. He told the crew to rent a large U-Haul truck, instructed the Jungle Boys to wear army fatigues, and said that this time, every one of them should wear ski masks as well. He loaded the truck with twenty guys and additional Jungle Boys followed in cars. They backed up the U-Haul outside Oasis at 1 A.M. as the club was closing. As their drunken enemies poured out of the establishment, Outlaw opened the metal door of the truck and out leapt twenty Jungle Boys, armed to the teeth. They were joined by the additional crew who had followed in cars.

Outlaw stood in front of his small army. "Next time any of y'all tell us where we can go, where we can't go, you're all gonna die."

Astounded, the clubgoers said nothing. After that, the Jungle Boys went to Oasis whenever they wished.

OTHER SKIRMISHES THAT summer, however, did not end so well.

One of the Jungle Boys had a girlfriend in college in Baltimore. Select members of the crew would occasionally go to the campus, eat in the dining hall, and play basketball in the gym. They stayed in a Sheraton and went sneaker shopping at a nearby mall. With their ostentatious garb, Outlaw and the Jungle Boys attracted attention wherever they went. Some of the Jungle Boys slept with the coeds after drunken campus parties.

On the way home, they would often drive through East Baltimore, a particularly malevolent part of town, where they would buy weed or hang out in a pool hall and get something to eat. On one such trip Outlaw and Ricky were walking to their car, after having eaten at a popular barbecue place. A couple of boys in the crowd noticed Outlaw's and Ricky's gold chains and asked if they could take a look at them. Their tone was clearly aggressive and provocative. Ricky and Outlaw soon found themselves surrounded by ten or so young men—it was hard to tell exactly how many in the poorly lit parking lot—who seemed to be edging for a fight. It was going to be ten against two.

Outlaw said, "I don't know what you're looking for, but you got the wrong guys. We just passing through." At the same time, he felt for his gun.

Out of nowhere—for who could see anything at all in the dark?—shots went off and Outlaw felt strange sensations in his body. A burning heat seemed to be emanating from his leg—his shin maybe?—and another from the side of his stomach. He felt his shirt getting wet. He almost lost his footing. The crowd backed away. Fortunately, Outlaw and

Ricky were right next to their car and Outlaw fell into the passenger seat, while Ricky took the wheel. The chaotic, milling crowd stepped back as Ricky drove away down the Baltimore street.

In the car, Outlaw moaned. The burning in his leg had reached a searing intensity. His stomach was throbbing with a white-hot heat. *I can't believe I've been shot,* he thought. The seat under him and the car floor at his feet were pooling with blood. The neon signs and streetlights of East Baltimore flickered in and out of his vision. At some point, Outlaw seemed to perceive that Ricky was stopping at a convenience store and at front stoops, frantically asking people questions. Outlaw wondered what he was doing.

At last, Ricky pulled up to a murky house and helped Outlaw up the stairs and into a room. A man was there and he told Outlaw to get onto a clean, white bed. He gave Outlaw a pill and then the man took out scissors and metal instruments. Some time went by and Outlaw felt the pain take over everything and then fade away. When he returned to some kind of consciousness, he seemed to think the man was stitching him up. The pain had subsided some, and Outlaw was brought back down the stairs and to the car by Ricky, who drove him back to New Haven. Ricky got him a bottle of whiskey and Outlaw fell out on the way home. In New Haven, Ricky brought him antibiotics from God knows where and creams and bandages and everything else that Outlaw needed.

As Outlaw came to understand the incident, it appeared that he'd been shot along the side of his stomach, the bullet passing straight through but not impacting any vital organs. The shot in his leg had hit the fleshiest part of his calf—which was of course far fleshier than an average person's calf—and that bullet had been deemed irremovable by the strange man in the strange house in East Baltimore. Outlaw's Jungle girlfriends changed the bandages daily and applied hydrogen peroxide. The bullet was still inside his leg, but the wound was stable and clean.

Outlaw asked Ricky, "How the hell did you know where to take me?"

Ricky replied that he had asked around for the location of a bootleg doctor. Ricky knew he couldn't bring Outlaw to a hospital, as the police would then get involved. It took a while but finally someone on a stoop had told Ricky exactly who to go to, probably some former doctor who had lost his license. It was the kind of information that people in East Baltimore knew about. Ricky paid the man $3,000.

Within a week, Outlaw was back on the hill, directing the Jungle Boys.

BUT IT WOULD not be for long. A particularly unstable boy called Nelson had worked briefly for the Jungle Boys and run off with $5,000. Outlaw put out word to track Nelson down. Not long afterward, Outlaw, accompanied by other Jungle Boys, ran into Nelson at a Kentucky Fried Chicken on Dixwell. They chased after him for three blocks down a side street. Nelson disappeared ahead of them, then popped out from behind a tree and began firing at Outlaw. He was clearly targeting Juneboy; no other members of the crew were fired upon. One of the bullets hit Outlaw in the head. Bleeding profusely, and gasping for air, Outlaw was brought by the Jungle Boys to the Yale New Haven emergency room, where he was anesthetized.

He was exceptionally lucky, the doctors said. The bullet in his head was lodged in the bone of his lower skull. The odd thing was that the bullet was perfectly stable there, and had not hit any brain tissue. It would be too risky to operate, the doctors said; they were going to have to let the bullet be. Pearl visited and prayed, and it all somehow seemed to work: Outlaw, perhaps because he was young and strong and determined, recovered remarkably quickly, even if the bullet is still lodged in the base of his skull to this day.

THE MAJOR EPISODES of the summer began in early June and involved a young man named Fly Boy, who first came to Outlaw's attention when he wanted to join one of the Jungle Boys' satellite business sites on Winchester Avenue. Outlaw didn't really know Fly Boy and deferred the hiring decision to his local crew.

"Do you want to put him down?" he asked them.

"No," his guys said. They said that Fly Boy was too impulsive and edgy. The next time Outlaw was over at Winchester Avenue, Outlaw told Fly Boy, "Look, we don't need you right now. But here's two hundred dollars." Outlaw could see that Fly Boy needed the money.

A week later Outlaw got a call that Fly Boy had pulled a gun on a Jungle Boy in the bathroom of Oasis. Sighing, Outlaw, who was on his way to a party in New York, drove to the club. It was closing time and the patrons were spilling out onto the street, mainly drunk. One of them was Fly Boy.

"What the fuck you doing? Out of respect I fucking gave you two hundred dollars," Outlaw said.

Fly Boy responded with a punch to Outlaw's face. Outlaw punched back. A crowd gathered around, cheering them on, and the sloppy exchanges quickly devolved into a wrestling match. Fly Boy, who was strong, clenched Outlaw in a headlock. He held Outlaw for some time in a stalemate, until the lights of a squad car lit up their bodies. It took three cops to separate them, and once Outlaw and Fly Boy were extricated, the cops arrested Outlaw for assault and let Fly Boy go. Clearly the police were much more interested in apprehending William Outlaw than Fly Boy. Outlaw was brought in a paddy wagon to the New Haven jail, where he quickly bailed himself out.

Three weeks later, on June 29, 1988—two days after the Mike Tyson fight in Atlantic City—Outlaw was on one of his frequent respite visits to Pearl's house. He was asleep in his old bedroom, when he was awakened in the middle of the night by his mother.

"Juneboy, somebody threatening you out there on the street. Saying ugly stuff he gonna do to you," Pearl said.

"Who is it?" Outlaw asked.

"I don't know, but don't go back out there, Juneboy. Don't do it!"

Outlaw peered out the window and saw Fly Boy on the sidewalk. Outlaw went outside and screamed at him, "Don't you ever threaten me or my mother. I can't get you right now out of respect for my mother, but I'll be coming for you."

Fly Boy rode away on a bicycle.

Back in the townhouse, Outlaw called the Jungle Boys. "Bring some guns, bulletproof vests, and ski masks. We're on."

"Don't go back out there. Don't go back out there!" Pearl shouted again. "Ma, he's threatening me at my home. He threatened me around you. I don't stand for that."

The Jungle Boys pulled up ten minutes later, guns in hand. In the last few weeks, Outlaw had learned where Fly Boy lived. He told the crew, "We going through the side alley by his house, hop the fence, and catch the motherfucker by surprise."

Minutes later, Outlaw and the crew found Fly Boy sitting almost innocently on his front porch. He didn't even have his gun ready. Outlaw, incensed that Fly Boy had threatened him and his mother multiple times, couldn't resist lifting his ski mask so Fly Boy could see his face before he shot him. "Why'd you disrespect me?" Outlaw shouted. Not waiting for an answer, Outlaw hit Fly Boy up with five shots. The light on the porch was dim but Outlaw could see the bullets tearing into Fly Boy's clothes, and Fly Boy falling to the porch floor. Outlaw was pretty sure he was dead.

The crew fled back to the Jungle and turned on the police scanners: "Fly Boy shot in Newhallville. William Juneboy Outlaw, armed and dangerous, chief suspect."

Outlaw thought, *Damn, this motherfucker must still be alive, if he's identifying me, because there were no witnesses around when we shot him.* Indeed, the police had found Fly Boy shortly afterward in the gutter. He was bleeding profusely and told the police, "If I die, Juneboy did it."

Fly Boy was brought to the hospital and Outlaw was arrested that night. Bail was set at $250,000. But then something strange happened. Fly Boy, who recovered with astonishing speed, didn't show up for the probable cause hearing. The charges were temporarily dropped, and Outlaw was free to go home.

But "free" was a relative term. Less than three months later, on September 24, 1988, Sterling Williams and Fitzroy Phillip, the two members of the Jamaican posses the Rats and the Cats, walked into the Jungle to kill the king. William Outlaw shot Sterling Williams in the head, killing him, and then hit Fitzroy Phillip multiple times as Phillip ran away.

THE ARREST WARRANT application for William Outlaw completed by Detective Joseph Pettola on September 24, 1988, stated the following:

> Officer John Bashta was in the vicinity of Spring Street and Liberty Street when he heard approximately five (5) to six (6) gunshots. Officer Bashta transmitted this over his radio . . . the New Haven Police Department frequency #3 dispatcher transmitted that there was a report that a person had been shot; laying on the ground at the location of Church Street South and Station Court, on top of the hill. The Officers Bashta and Timothy responded to this location and found a black male laying in the street suffering from a gunshot wound to the head . . . it appeared that this black male was lifeless.

The warrant then went on to describe how Fitzroy Phillip was found wounded one block away, brought to the hospital and treated for his wounds, and subsequently testified:

> [Phillip] described the individual that shot him and Williams as a black male, approximately 6' 3" tall, approximately 250 to 260 pounds, dark complexion, short hair, and attired in all white outer garments . . . I was informed by Officer Bashta and Detective Burton that a black male named William Outlaw nick name "June Boy" who matches the physical description was seen earlier in the night . . . at the location of Church Street South and Station Court attired in all white clothing.

In a subsequent document, Pettola wrote:

> On Sunday, September 25, 1988, an autopsy was performed at the Medical Center in Farmington by Dr. Luke on the remains of Mr. Sterling Williams. The death of Mr. Williams was ruled a homicide

and during said autopsy the cause of death had been established that the gunshot wound to his head caused Mr. Williams' death.

WHEREFORE, the undersigned states that he has probable cause to believe that the said William Outlaw did commit the crime of Murder, in violation of Section 53a-54a of the Connecticut General Statutes.

Pettola wrote and signed a second arrest warrant for Outlaw, for two counts of assault in the first degree (firearm) in the shooting of Fitzroy Phillip.

As Pettola was busy interviewing police and witnesses and writing the warrants, and while the New Haven police put out an all-points bulletin to find William Outlaw, Outlaw stayed at the Novotel in Midtown Manhattan while he figured out his next move.

In earlier skirmishes with the law, Outlaw had asked the bail bondsmen who was the best defense lawyer in town. They consistently named one man: Ira Grudberg. Indeed, Grudberg was widely considered the best defense lawyer in New Haven, if not the state. He finished sixth in his class at Yale Law School and had been practicing for thirty years. Grudberg was brilliant and flamboyant, known for wearing snakeskin belts and frequenting the bars downtown. Outlaw had retained Grudberg to defend the Jungle Boys on drug possession and sales charges, and Grudberg had done an excellent job. Outlaw and Grudberg—who was five foot eight and Jewish—were an odd couple but they developed an easy rapport. Grudberg didn't share his personal background with Outlaw, but he too had grown up poor, in the projects in Bridgeport where his father had been sporadically employed as a deliveryman for an "egg and butter" operation. Grudberg delighted in representing the underdog, or as he put it, "taking on the man, and knocking him down a little bit." But Grudberg didn't come cheap. To represent him on the murder and assault charges, Outlaw paid Grudberg $160,000 in cash.

Outlaw called Grudberg from his Manhattan hotel. The lawyer advised Outlaw that the case against him had the potential to be argued as flimsy because of the chaos at the murder scene. It could be hard to ascertain who exactly the killer of Sterling Williams was. But Grudberg also told Outlaw that eluding the arrest warrants would only create more problems and reflect his possible culpability. Outlaw needed to return to New Haven as soon as possible, turn himself in, and plead not guilty. On September 29, Outlaw arrived at Grudberg's office at 7 A.M. Together they drove to the police station in Grudberg's Honda Accord. (Even though he was making a million dollars a year at that point in his career, Grudberg remained frugal.) The lawyer told Outlaw to remain silent as he was booked, processed, and charged with murder and assault in the first degree.

Bail was set at $1.5 million. Outlaw was originally supposed to be held in the New Haven jail until the court date, but he had an older cousin, whom he barely knew who worked as a corrections officer there. The court decreed it was a conflict of interest for Outlaw to be held in New Haven. He was transferred to Bridgeport jail, where he was held for nine months awaiting trial. But given his previous institutional experience at Long Lane and Manson, Outlaw didn't find the experience particularly onerous. On his first night, a guard left a takeout pizza on his bed. Outlaw had a steady stream of visitors: Camile, the many other girlfriends, many Jungle Boys, and Ira Grudberg, with whom he plotted strategy during visits on Sunday nights. Outlaw was not terribly concerned about being found guilty, and was eager for the trial to begin.

The trial *State v. Outlaw* finally commenced in late July 1989, in the New Haven courthouse, a handsome and imposing 1917 beaux arts structure in the heart of downtown, overlooking the Green. The courtroom had high ceilings, oak-paneled walls, benches that were stained a rich chocolate color, and a thick royal-blue carpet. The room exuded stoutness and venerability: a place in which justice was supposed to be done.

The prosecutor was a young and affable attorney, Patrick Clifford. Clifford, in his mid-thirties, was good-looking and blessed with an ease of manner and the Irish-American gift of gab. Everybody liked Patrick Clifford, even his opponents. Clifford was even friendly with Outlaw, whom, in Clifford's typically amiable style, he called "Willie." Outlaw had been in and out of the courts so often in the last few years, as both victim and perpetrator, that Clifford had gotten to know him well. But underneath his easygoing frat-boy demeanor, there was a hard-driving aspect to Clifford. He desperately wanted to beat Ira Grudberg in the case. Grudberg was a decade and a half Clifford's senior and had amassed a near-legendary reputation in the city's legal community. A victory would be a significant feather in Clifford's professional cap. But Clifford had a major problem with the case. He didn't have a witness. Fitzroy Phillip, after delivering his bedside testimony at the hospital and appearing at a pretrial probable cause hearing conducted by Clifford a few weeks after the shooting, had simply disappeared. Clifford had sent members of his staff to Brooklyn and they had spent days trying to find him. They had a home address from the interview and when they arrived at the apartment building they saw a man who resembled Phillip. Clifford's assistants approached the man, but he ran away. Without a witness, Clifford could rely only on Phillip's testimony from the preliminary hearing.

Grudberg would not let Outlaw take the stand, nor would he let Jungle Boys attend the trial. Only Pearl, members of the extended Outlaw family, and the mothers of Outlaw's growing number of children—he now had five offspring from various women—could attend. Camile Leslie chose not to attend the trial and spent her time praying at Pearl's Mount Olive Church. When the trial began, Grudberg implored Outlaw not to smile, lest the members of the jury see his gold-plated teeth. Outlaw ignored him, beaming broadly at the jury throughout the proceedings.

Without Fitzroy Phillip as a witness, the police officers' testimony was the cornerstone of the prosecution's case. On the stand, Officers Gilbert Burton and Joseph Pettola explained how Fitzroy Phillip, even in his compromised state in the hospital, had clearly and lucidly identified Outlaw as the shooter of both himself and Sterling Williams. Grudberg retorted that "this case is innuendo and guessing." Grudberg addressed the jury. The summary of his remarks was: "You do know that Williams was killed and Phillip was shot. But there is no real evidence here, no prints were taken at the scene. There's no proof that Outlaw even had a gun. I have never had a chance to cross-examine Phillip and so you are asked to judge the testimony without the person being present. There is no way to assess the credibility of what he claims. The State is not at fault, but Phillip is 'lost' to these proceedings." Clifford read Phillip's pretrial testimony in the courtroom, and in brief closing remarks implored the jury to "use your common sense. What the defense says is innuendo, the State thinks is common sense. Outlaw is huge, and very easy to remember. When someone that size shoots your friend, you're likely to recall it."

The jury did not deliberate for long. On July 28, 1988, the verdict that was returned by the jury was unanimous: Outlaw was guilty of murder and attempted murder. The eventual combined sentence was sixty years. Outlaw was to be sent to Somers Correctional Institution, the only maximum-security prison in the state, immediately.

Outlaw was stoic as the verdict was read. Before he could really process anything, he was handcuffed and led by court personnel to two waiting state police cars outside. Outlaw didn't even have time to say goodbye to his mother. He only heard her gasp at the verdict.

Minutes afterward, he was alone in the back of a Ford Crown Victoria police cruiser, the sirens on and lights flashing. The car drove eighty miles an hour all the way to Somers. It was as if they couldn't get Outlaw to prison—and out of New Haven—fast enough.

CHAPTER 4

Release Date: April 7, 2073

Somers prison, 1990.

LATER THAT DAY, NEWLY MINTED CONNECTICUT STATE INMATE #156049 had only half an hour previously been locked inside his cell—a five-by-nine-foot space, with a metal bed, a toilet and sink, cinder-block walls, and a front door made of tightly spaced gunmetal bars—when a corrections officer stood outside, saying that he had a letter for him.

"How the fuck could I have gotten a letter, motherfucker! I've only just gotten here," Outlaw seethed.

He wanted only to be left alone at this point. He had been in a state of disbelief since his sentencing earlier that day. The moment the word "guilty" issued from the jury foreman's lips, the terrifying weight of it all—at once profoundly real and absolutely unfathomable—had not lifted for one millisecond. With that one pronouncement, William Outlaw, the great and untouchable gangster, had been shunted into a new universe. In the blurred moments immediately after the verdict, Outlaw had been descended upon by armed officers like a captured animal, shoved into a squad car, driven the seventy miles to Somers prison at top speed, been subjected to a soul-crushing intake process in which he had variously been weighed, fingerprinted, photographed, X-rayed, searched in all bodily cavities, issued a white T-shirt and khaki pants, and been given all kinds of meaningless documents to sign. He hadn't bothered to read any of them. But now, mercifully, he was alone in his cell and had just experienced the first moment of peace—a highly relative term, of course, given the circumstances—in hours. This most horrible of days was finally over. But no, here was a guard in his navy-blue uniform, with an American flag and Connecticut Department of Correction patches emblazoned on each shoulder, with a letter for him.

The guard slid the envelope through the bars. Outlaw opened the seal, pulled out the letter, and saw that it was on official Department of Correction stationery and addressed to "Inmate William Outlaw." Below was his "date of admission to DOC facility," July 28, 1988. Below that was another date, "anticipated date of release," July 2048, reflecting the sixty-year sentence he was issued earlier that day. Somehow, in the chaos of the courtroom, he had failed to take in the length of his punishment. After the verdict, his primary concern had been for Pearl, sitting in the row behind him. He had turned around to embrace her, but the guards had arrived so quickly that he barely caught a glimpse of her anguished face before he was marched off. Now here it was in black and white: July 2048. Outlaw's stomach felt like it had dropped twenty stories. He clasped the bars of the cell door to keep himself from falling. He thought, *I am going to die in here. I am going to leave in a casket.* Outlaw didn't know how long he clung to those bars—it could have been seconds or minutes or even hours, because time no longer mattered—but eventually his grip weakened and he fell in a great heap onto the floor. He lay in a kind of fetal position. Outlaw didn't sob—he couldn't allow himself to, for that would attract the attention of inmates in adjoining cells and announce that he was weak. Instead his mind seemed to depart his body, floating up through the ceiling of first the cell, and then the facility, and into the sky, where it could be untethered and free. With his mind up there—out *there*—in suspension of all reality, he looked back down to earth at his twenty-year-old self, still unable to move on the floor of the cell where he would remain for the next sixty years. At some point he said a prayer. He didn't ask God for anything—he had never believed in that kind of pointless beseeching—but he made a promise: *If I ever get out of here, God, I promise I'll do something with my life. I don't know what, but I'll do something . . . something better.* He lay there all night, and he must have slept at some point because in the morning a guard roused him. For a brilliant moment he forgot what had happened the day before, but then it all came crashing down upon him again.

The guard said, "Inmate Outlaw, I'm not supposed to do this, because we're not authorized to let you see anyone yet, but there's someone here to see you." The guard brought Outlaw to the visiting room, which, with its cheap desks and chairs bolted to the floor, vending machines and small barred windows, was just as depressing as the rest of the prison. It was so early that there was only one visitor, a frail-looking middle-aged woman in a plain gray cloth coat, sitting at a table. At first Outlaw didn't recognize his own mother. Pearl was wearing makeup, looking oddly dressed up for the occasion. Outlaw sat opposite her, under the watch of a guard in a bulletproof Plexiglas booth.

"I needed to tell you that last night I slept the best I have for years, knowing that you were in a safe place," Pearl said. "God told me I had to come here to tell you that."

Outlaw nodded.

"God also wanted me to ask you, did you kill that man?"

Outlaw said, "God knows everything, Ma. He knows what happened. I don't have to tell you."

She looked accusingly back at him. "Tell me."

"God knows, Ma. Only God knows."

Pearl was not impressed by his answer. Then Outlaw added, "Okay, Ma, I'll tell you if I killed that man if you tell me one thing."

"What's that?"

"How did you get that long scar on your face? Who did that to you?"

"I'm not going to tell you that."

"Then I'm not going to tell you if I killed him."

Pearl grabbed her purse and got up to leave. The visit had lasted five minutes from beginning to end. It was one of Pearl's last visits to Somers. In her mind, she had worked too hard to raise a felon. She had done everything she could, and now maybe only the system's punishing ways could save her son.

Over the rest of the weekend, Outlaw was sequestered in his cell, unable to go to the yard for rec or the mess hall. His meals, which were predictably odious, were delivered on metal trays. From his cell, he heard catcalls from shadowy figures in the cells on the other side of the tier. "Yo, Outlaw. Welcome to the Jungle!" "Juneboy is here!" "What up, JB?" It seemed like the whole prison already knew that Juneboy Outlaw had arrived.

Early Monday morning, Outlaw was told to meet with a Captain Brunson, a senior corrections officer. Outlaw could tell from the way that Brunson looked him straight in the eye and spoke without prevarication that he was that rare thing: a person in authority who was respectful and honest. Outlaw was respectful in return. On Brunson's desk were twenty notes written on slips of yellow paper, and he showed each one to Outlaw. The messages variously said: "I will kill Outlaw." "Fuck him." "I will murder Juneboy the moment he comes on the unit." Brunson explained that the notes had been left by inmates in collection boxes on the block through which they could make their opinions known to the administration.

Brunson went on, "I strongly recommended that you be housed in the Ad Seg unit, which stands for 'administrative segregation.'" Outlaw knew what Ad Seg meant. All real criminals did. Ad Seg was reserved for prisoners with vulnerabilities such as mental illnesses and chronic diseases or those who required special protection because of the particular heinousness of their crimes, such as child molestation and rape.

Outlaw smiled at Brunson and said, "I'll take my chances with gen pop." It was unthinkable that he would ask for any form of special protections.

"You sure?"

"Yeah I'm sure!" Outlaw laughed. He would be the joke of New Haven: Juneboy in Ad Seg!

"Okay, but you will need to sign this form indicating that you are taking full responsibility for your own safety," Brunson said.

"Sure," Outlaw agreed.

In his cell on the general population unit later that day, Outlaw surveyed the minuscule space that was going to be his home for the next sixty years. The only redeeming feature was that he had a private cell. Virtually all prisoners lived with a bunkmate, or "cellie." But presumably because of Outlaw's size, the concern for his safety, and what turned out to be the well-founded fear of the deleterious influence he would have on the institution, Outlaw was housed alone, and indeed would remain so for the entirety of his prison career. Outlaw peered out of the cell's one comparative luxury: a small barred window overlooking the prison yard, itself a large and unadorned enclosure surrounded by a concrete wall with wired fencing on top and gun towers at its corners. He heard the moans and talk and wails of other prisoners. As Brunson had warned, some of the chatter was directed at him. "Suck my dick, Outlaw!" "I'm going to fucking cut your head off!" "I'm going to shank you!"

He said nothing in response. Accustomed as he was to people coming after him with guns, he didn't even particularly consider the statements as threats. *We'll see if they say the same thing when I'm out on the yard with them,* he thought. He also couldn't help but take some pride in the fact that his reputation, as well as that of his family, had preceded him. Indeed in the coming weeks, older guards spoke to him about his grandfather. "You have an unusual name, Outlaw. You related to another William Outlaw that was here a while back? He was a decent guy, a quiet guy. If you're anything like him, you'll be okay here."

Outlaw didn't plan to be anything like his grandfather during his time at Somers.

From the moment that Outlaw arrived at Somers, his desire was to punish the system. He had roamed wherever he wanted up and down the East Coast, and now he was a caged animal. It made him insane. He had been screwed over. His territory in the Jungle had been invaded by foreigners and he had simply defended it. He had single-handedly kept the Jamaican posses, perhaps the most dangerous gangs in the world, out of New Haven. In his view, he had performed a vital community service, and was being punished for it. The only thing that made him feel better was knowing that he might be able to inflict significantly more damage in prison: on officers, on the prison, on the system, on anything or anybody that got in his way. He was going to maraud the place, make his mark on it so that it would never quite recover, so that for years people would talk about the impact of William Juneboy Outlaw III. If he could burn the place down, he would. With a sixty-year sentence, what was there to lose?

He observed that much of the actual operations of the facility, and therefore a portion of its power, was in the hands of its residents. Inmates staffed the gym, the chow hall, and the library; they took many of the routine phone calls and made most of the food. Outlaw was surprised at this, and it made him consider the opportunities that lurked therein, for this meant that inmates had access to closets in the woodshop and storage containers in the gym, ideal places to keep and hide things. What he needed to do was study this place and see how it worked, so that he could eventually identify its weak spots. Because every human system, he knew, had its vulnerabilities. The organizational hierarchy, Outlaw saw, was that of a warden, a deputy warden of operations, various lieutenants, numerous unit managers and captains, and finally, corrections officers and counselors. The institution revolved around a complex schedule of counts, meals, rec times, work programs, and educational programs, all running on a staggered schedule to accommodate the overflow of prisoners. He noticed who worked on which shifts, who officially ran the place, and who unofficially did.

Outlaw refused all services that were offered to him, such as classes toward a high school equivalency degree, support groups and counseling of various kinds, as well as opportunities to work in the kitchen or the library. He was not in prison to get help. He didn't need any help. "Fuck you and your motherfucking programs," he said to the counselors. He had a particular disdain for the prison industries, which included making license plates at fifty cents an hour. "I'm not gonna be your nigger," he would yell at the staff. When he learned that inmates were allowed to wear necklaces for religious reasons, Outlaw had one of the crew in New Haven slip him a $5,000 gold chain during visiting hours. The chain had a Buddha attached to it and Outlaw told officers that the pendant around his neck was a testament to his deep and abiding faith in world religions.

He began a prolific career of accumulating disciplinary tickets. The following complaints by staff were typical: *Inmate Outlaw was observed loitering around G-Block for over 10 minutes. When he started walking back toward his block he made the following comments: "I got 60 years, asshole, fuck your mother with the ticket." Approx. 70 other inmates heard this and stopped to watch.* Or: *Inmate called Captain Perry "the Head House nigger" . . . telling us to "suck his dick."* Or: *Outlaw invited me into the closet stating he would kick my ass. Stated "it won't be long I fuck all of you up."* Or: *Inmate was locked up in the interview room, on the way there he "motherfuck"ed me till he was locked up.*

In his search for vulnerabilities in the system, Outlaw happened upon one early on in his prison career. In a corner of his unit was a bank of pay phones. Outlaw would get a stack of quarters and call Pearl, who was still unusually frosty, and Camile. As he listened to other inmates making calls, he was astonished to hear them talking freely about criminal behavior. Sometimes they used code words for drugs and violence, and if a guard happened to walk by they would switch topics, but it was clear what was being discussed. It also appeared that no one was monitoring these calls. And so Outlaw began to call the Jungle Boys—Rodney, Alfalfa, Pong—and told them to be on the lookout, that he might be able to get some drug dealing going soon within the prison confines.

On the yard and in chow hall, Outlaw was sure to project the right measure of smoldering anger and malevolence, indicating to everyone around him he could attack at any time. In many ways, it was simply an act of self-protection. During his first week, Outlaw was looking across the tier from his cell and saw three inmates entering the bathroom. Shortly afterward, he saw two of the men walk quickly out. Then the third inmate stumbled out into the hallway clutching his neck, blood spilling from it. He cried for help. The guards showed up without much urgency and took the man away in a wheelchair. A few weeks later, Outlaw was in the shower by himself when he heard sounds coming from a stall. Two inmates were raping a third. From that point on, Outlaw never closed his eyes in the shower room.

At lunch in his first week, two Jamaicans came to his table. "We are part of the posses, the Rats and the Cats," they said. "We know all about you, Juneboy. We knew Sterling Williams and Fitzroy Phillip. They were our friends. And you know what we heard about you? We heard you like to kill Jamaicans for sport."

Outlaw said nothing. He needed to think before he said anything.

"We're here to tell you we not gonna stand for dat. Fuck with us and we'll kill you," the Jamaicans said.

"No, no, you guys don't understand. The dude came into my turf. He was armed and coked up," Outlaw said. "He was coming after me and my gang. I don't got nothing against no Jamaicans. In fact, I respect you for coming to me directly to tell me your beef."

Outlaw told them the story of what happened on the night that Williams entered Church Street South. He supplied specific details—time, place, weather—things that he couldn't have possibly made up on the spot. The two Jamaicans listened to the long, involved story. "Okay," they said, and got up to go to the yard. They shook his hand as they left.

THE PRISON AT Somers to which Outlaw arrived was overrun by hardcore criminals: drug dealers, gangleaders, murderers, rapists, and other seasoned offenders of all kinds. Mike Lajoie, who began his career there as a corrections officer, observed, "Somers in the 1980s and 1990s was one of the most violent and hostile prisons in the entire country. It was like something out of the movies." Somers was constructed in 1966 to house 900 inmates, but with the recent influx of gang members, the census was now 1,400.

By the time Outlaw arrived, the prison was in many ways beyond the control of the administration. The system seemed fundamentally ill equipped to deal with both the volume and the level of dereliction of the new inmates. It was as if Connecticut wanted desperately to cling to its own image and still couldn't quite believe that it had a gang problem despite overwhelming evidence to the contrary. In an attempt to manage the new generation of younger and violent offenders, the state hastily built new warehouse-like prison facilities, mainly in the remote and poorer northern regions of the state that didn't have the political clout to keep them out. In the 1980s, the number of prisoners in Connecticut doubled, the prison staff tripled, and the Department of Correction's budget quadrupled. For many communities in northern Connecticut, the new facilities were a godsend. The former agricultural economy was dwindling, and the prisons offered well-paying state jobs. Employment as a guard was considered a hazardous occupation and one could retire with a hefty pension after only twenty years.

It was common for prisoners to throw their feces at guards. Three guards had their throats cut during a riot and the warden was assaulted while touring the woodshop. The year after Outlaw arrived, inmate George Guedes was set on fire in his cell by two members of the Latin Kings, a Hispanic gang formed at Somers. The first Latin King sprayed paint thinner, likely acquired in the prison workshop, out of a shampoo bottle into Guedes's cell. The second inmate, following the first, threw a pack of lit matches onto Guedes. The incident was caught by video camera in the hallway. When prison officials watched the tape later, they saw the two men executing their plan in perfectly choreographed motions, followed by a burst of flames shooting out of the cell. Guedes died two days later. Mike Lajoie, the young prison guard, said you could smell burning flesh for four days. "It turned your stomach. The funny thing about it was that they were not even violent inmates who did it," Lajoie noted. "They were what we called 'nothing inmates.' It was the higher-ups who told them to do it. That is the way it usually worked: the real heavy hitters are untouchable, and you have these little guys who actually do the stuff and get caught for it." Outlaw too smelled the stench from his cell and knew immediately that an inmate had been burned. There had been a dog pound not far from Q View, and he could always tell when animals had been incinerated from the otherworldly odor that pervaded the neighborhood.

Most of the staff were white, country boys from the rural towns that surrounded Somers. Many guards lived in Massachusetts, the border of which was just seven miles from the prison. These villages, generally with one to three thousand residents, were often called "cow towns" and still clung to the remnants of their agricultural pasts. Amid the modest ranch houses and slightly more upscale housing developments, there were still the occasional dairy and crop farms. In many such towns, the entire commercial enterprise might be a gas station, a 7-Eleven, and a pharmacy that sold everything from rock salt to videotapes. The extent of the education among most of the Somers guards was a high school diploma; many of them were big hulking guys, former football and baseball players working on a paunch. They typically drove pickup trucks, or if they had been on staff sufficiently long, revved-up Mustangs or Camaros, from which they blasted Bruce Springsteen, Lynyrd Skynyrd, Tom Petty, and John Cougar Mellencamp—straight-ahead rock and roll with a country twang. Many of them had done stints in the military and served in the National Guard on weekends. Virtually all of them were fans of Ronald Reagan.

To Outlaw, the guards were simply racist assholes. The corrections officers were, in short, the enemy—another, if more rural, iteration of the hated "blue and white" from New Haven. Outlaw found many of them were ridiculously easy to intimidate. It was evident that many of the guards, under their uniforms, were petrified. They were green, just out of high school, and unused to black people or anyone from an urban environment. And, of course, Outlaw was physically larger than every single corrections officer on staff. Outlaw made a game of it, like a cat toying with a dazed mouse. He would call a corrections officer over and ask where his wife and kids lived. Or ask him why they couldn't get a better job than taking care of a bunch of niggers. What could they even do to him, anyway? Put him in Ad Seg, or the solitary unit? He'd always wanted to see what solitary was like. As far as Outlaw was concerned, this was his house, and the guards were just babysitters.

After violent incidents, the facility was placed on lockdown, meaning that staff would be unable to leave once their shifts were over. Often they had to stay on for sixteen or even twenty-four hours of overtime. Inmates would be sequestered in their cells until the authorities completed their investigations. Outlaw was deeply suspicious of lockdowns. He thought they were intrinsically in the prison staff's and prison union's interest, creating bounties of overtime pay and costing the taxpayers millions of extra dollars. In Outlaw's view, everyone was just getting rich off people like him.

There was, however, one group of corrections officers, a subset of the guards, who were up to the challenge of inmates like Outlaw. In fact, they made it their explicit goal to control the institution and get the most incorrigible inmates to conform. They called themselves the Northern Boys. There were about fifteen of them in all, and they were amped up with the same level of testosterone as the hardest inmates. The Northern Boys prided themselves on staying in military-level physical shape. On the weekends, they played football, fished, and hunted for deer. At the center of the Northern Boys was the young corrections officer Mike Lajoie, who at age twenty-four brimmed with confidence and had the energy and ambition to match Outlaw. No matter how late Lajoie worked—and he worked many extra shifts because he was so effective at his job—he would be at the gym at five the next morning lifting weights. At five foot nine, he wasn't especially tall but he was powerful and had the tightly coordinated movements of a champion wrestler. His face was chiseled—everything about him was angular and hard. Outlaw and Lajoie were on opposite sides of the war, but they shared many similarities, and their fates would intertwine over the course of the next three decades in ways that neither would ever have predicted.

Mike Lajoie began his career out of high school, patrolling state parks for the Department of Environmental Protection. But he discovered that not much happened in state parks other than fishing and hiking and teenagers smoking pot. Lajoie soon transferred to the Department of Correction, where he found his niche. Even though he had grown up in the desolate rolling hills of New England, he was naturally street smart. He had an easy rapport, one that was both tough and fair, with inmates, and had little tolerance for wrongdoing. Lajoie's primary tactic in combating the hard-core inmates was to learn to think like a criminal himself. He tried to put himself in their shoes, and then keep one step ahead. If he was trying to get drugs into the place, how would he do it? If he had little social support and little to lose, how would he behave? By thinking this way, Lajoie realized that anger was the default emotion in Somers, and that the anger was used to smother the second emotion that drove the place, fear. Lajoie understood how humiliated many of the inmates were underneath their swagger. Given their tenuous self-esteem, he saw how critical their perceptions of "respect" were, and why they would take such offense if someone looked at them oddly or talked to them unprofessionally.

Lajoie spent a lot of his time observing Outlaw. He knew his reputation from New Haven, and he saw Outlaw gradually building his sphere of influence, hobnobbing with other inmates. He noticed how crowds were starting to grow around Outlaw in the rec yard.

Outlaw knew the first step in taking over the place was getting access to drugs. Access to pot and cocaine in prison was the key, just as it had been on the streets of New Haven. Drugs were already present in the facility—Outlaw could smell marijuana from cells and bathrooms—but their presence seemed infrequent and the supply piecemeal. Desperate to numb themselves to their surroundings, many inmates were reduced to smuggling fruit cups from chow and hiding them in their cells until the juice fermented into alcohol. Outlaw knew from his earlier experience at the Manson Youth Institution that the simplest way to get drugs into a facility was through direct exchanges with visitors. But this was risky and unpredictable. He didn't want to repeat his earlier mistake, which had helped turn an expected three-week stay at Manson into four months. He needed a more reliable route.

From his vantage point in his cell, or out in the yard, or in the chow hall, he began to watch the guards. He studied them as they passed through the unit on rounds. He looked for guards who appeared overwhelmed and intimidated, or who worked a lot of overtime shifts, indicating that they might need money to feed an addiction. In particular he looked for signs of a habit: sniffing, agitation, pale complexion, glassy eyes. During a lull one afternoon, Outlaw was lying in bed, when a youngish guard appeared outside the bars of his cell. The guard looked up and down the hallway, checking to see that no one else was around. Outlaw could see the fear in his eyes. He was familiar with the corrections officer, an eager, fresh-faced, all-American type, and had noticed that he'd been working a lot of overtime recently. Up close, Outlaw saw that the guard was looking worn, even haggard. "You need help with anything?" the guard whispered.

"What the fuck you talking about?" Outlaw responded, indignant. "Get the fuck away from me."

The guard came back a week later, repeating his offer.

"What the fuck do you mean?" Outlaw said.

"Anything, anything at all."

"Talk to me in a week," Outlaw snarled.

A week later, the guard showed up again. Outlaw told him to slip him his home address on a piece of paper. The guard lived in some hick town near Somers. Outlaw in turn gave the address to one of the Jungle Boys during a visit. A few days after that, Outlaw told the guard, "Listen up. Someone is going to deliver a box to your door. Don't you dare open the box. All you need to do is bring it to me, put it under my pillow. Once I get it, my guy will drop two thousand dollars in cash at your house."

A week later, Outlaw found a shoebox wrapped in brown paper under his pillow. He opened the box and inside was just what he had hoped for: half a pound of marijuana and half a kilo of cocaine. He used a shank he had acquired to rip open the side of his mattress and pushed the bags of drugs inside. Then he stayed up all night sewing the mattress closed with white thread and a needle that he had bought for just this purpose at the commissary. He did a meticulous job: a casual observer wouldn't notice anything awry. Now he had a mattress full of drugs. Everything was going to be different.

In subsequent days, he became a drug dealer again. That wonderful feeling of mastery and power returned. He sold the goods at incredibly inflated prices. A joint or half a gram commanded ten times what it would on the streets. But soon, Outlaw, not wanting to get his hands dirty, hired others to do the dealing. There was no shortage of inmates willing to do his bidding for a piece of the action. The black inmates from New Haven, many of them his mortal enemies on the streets, clamored to join in. Within just a matter of weeks, the prison was transformed. Entire units now reeked of marijuana. The way Outlaw looked at it, he elevated everybody's lifestyle. Now that he had the ability to make their lives tolerable again, he became immensely popular.

Camile Leslie visited with a saintly fervor and loyalty three or four times a week, making the seventy-mile trek from New Haven even though she was full-time on the night shift as a nurse at Yale New Haven Hospital. She also did private nursing in the home of a semiretired attorney who was frail and diabetic. He happened to be a friend of Ira Grudberg, and Camile asked him for legal advice about Outlaw's case. Camile in some ways seemed to be more devastated by the verdict than Outlaw. She had been at church praying when the verdict was handed down, and perhaps because she was not in the courtroom to witness the finality of the announcement, she seemed not to have been able to fully accept the news. She sat with Outlaw in the visiting room every couple of days. Camile would pray and cry, and they would hold hands when the guards weren't watching. Sometimes she would slide her hand across the table and press a crumpled-up bill into Outlaw's hand. It was remarkable how small you could make a hundred-dollar bill if you really wanted to. Outlaw used the cash to buy brand-name goods in the commissary.

Thanks to his conduit via the corrupt officer, Outlaw started getting more drugs into the system than he could effectively store in his mattress. Having heard that there was a master screwdriver that opened a variety of doors and panels in the facility, Outlaw spoke one night to a janitor who offered to sell him one for $500. After being paid, the custodian handed Outlaw the screwdriver and pointed at the ceiling just outside his cell door. Outlaw looked up to see a fixture housing fluorescent lights.

"It will open that," the janitor said.

At 3 A.M., when the guards were half asleep, Outlaw moved his bed next to the cell door, and standing on it, reached through the bars to unscrew the light fixture. He placed the drugs inside the housing and screwed the fixture back into place. From then on, the guards would shake down the whole unit but they could never find the source of the marijuana smell. The guards would come through dumbfounded and Outlaw would smile at them happily with his gold teeth as they walked by.

IN PRISON PARLANCE, a "shot caller" is the leader of many inmates. The position is usually arrived at through intimidation, cleverness, and entrepreneurial pluck. Shot callers, in effect, re-create gangs in prison, and use many of the same techniques they used on the streets. The Guedes murder was the in-prison version of a drive-by shooting.

When Outlaw arrived at Somers, the Hispanic prisoners had two shot callers, Pedro Milan and Nelson Millett, who acted as co-leaders of the Latin Kings. Grandiosely, they called themselves the Supreme Crown. Milan positioned the gang as an appeal to Hispanic pride, stating once in a prison interview, "We shall not be deterred in our quest to achieve our goals for our Latin community." Even from prison, Milan and Millett exerted an extraordinary influence on the cities of Connecticut, dividing the state into five regions and appointing commanders, presidents, and soldiers. Officials estimated that there were at least a thousand members of the Latin Kings, perhaps ten times that many. Milan and Millett appeared to have ordered multiple homicides.

For their part, many of the white prisoners were under the undisputed sway of just one individual, the ironically named Ronald "Tiny" Piskorski, who weighed in at three hundred pounds. Tiny's journey to Somers began on the night of October 19, 1974, when he, along with an accomplice, killed six employees and customers of the Donna Lee Bakery in New Britain, a blue-collar community ten miles south of Hartford. After a call from a concerned neighbor who was unsettled by the lack of activity at the usually buzzing establishment, a policeman entered the back room of the shop and found among the baking racks the bodies of six people, all of whom had been shot at close range. Two had been beaten severely with a hammer before they were shot. "I've never seen that much blood in my entire career," said the police photographer. The shop owner was lying facedown, the side of his head removed by a shotgun blast. The apparent motive was robbery: the victims' purses and wallets were gone, and the cash register emptied of $150. Altogether, the incident netted its perpetrators $300. It took three weeks, but the detectives' trail eventually led to Piskorski and his accomplice Gary Schrager, who were known as perpetual troublemakers, heavy drinkers and drug users, especially enamored with speed. One bartender said of them, "They were friendly when they were sober." Once, Piskorski entertained his fellow laborers at lunchtime by biting the head off a live snake. It came out in the trial that right after the shooting, Schrager and Piskorski went to the home of a friend. "I just shot six people," Tiny announced, sounding oddly happy. Piskorski was sentenced to a state-record six consecutive 24-to-life sentences. At Somers, Piskorksi, brilliant and charismatic, was known as a master wood craftsman in the shop.

Outlaw saw that there was no equivalent to Piskorski, Millett, and Milan among the black prisoners, and he decided it might as well be him. Who else could it be? With his reputation as a drug runner now established, Outlaw recruited lieutenants and various functionaries below him. His main second in command was Alvin Stewart, who was six foot four and 250 pounds and incarcerated for murder. Known as "Big Stew," he was committed and charismatic. Soon enough, Outlaw was walking down the corridor with three, four, or five guys surrounding him. In so many ways it was just like the Jungle Boys, but on the inside. Inmates getting stabbed and guards getting punched became regular, weekly occurrences. Lajoie knew that Outlaw and the other shot callers were behind it.

When inmates didn't pay their drug debts, Outlaw had one of the functionaries stab them with a shank or beat them up. Punishment came in specific forms: being punched, being slammed over the head with a bucket, being stabbed, having shit or piss thrown at them, being harassed, or having their families threatened. The going rate for a stabbing was $500 or $600. Outlaw had learned that in prison, stabbing was the best method. It left a scar for the entire prison to see as an ongoing reminder of the consequences of defiance.

Outlaw, having now acquired the power he once had on the streets of New Haven, decided to call his cell block the Terrordome. As with the term Jungle Boys, the name immediately caught on. The guards were put on notice that if they chose to truly attempt to enforce the administration's rules in the Terrordome, which was not necessarily advisable, they would be summarily punished by inmates if they did anything but look the other way. It got so that Outlaw smoked weed whenever he wanted, even blowing it in the guards' faces.

But Outlaw's master stroke at Somers was yet to come.

Another guard, one who also had a reputation for being particularly hardworking and abstemious, approached with the usual refrain: "Anything I can do for you?"

Outlaw was flabbergasted that this particular guard, with his stellar reputation, was making such an appeal. He responded with the usual sneer: "What the fuck you talking about?"

But the young guard kept coming back and each time Outlaw sent him packing. He needed to test the guard because this was going to be a major move.

After a month of rebuffing the guard, Outlaw finally agreed to work with him. He once again got the guard's home address and told him that a package would be arriving at his house at a specific time. "Do not open the package," he said, "or we'll hurt you. If the package is in any way tampered with, the deal's off and things will get physical."

In the visiting room, Outlaw told one of his crew members to bring a gun to the address that he'd just slipped to him.

"You sure, JB?"

"Yeah, I'm sure. Put it in a sneaker box and wrap it well. Wrap it, then double-wrap it. Then wrap it again. Bring it next Sunday at five P.M. to that address. Then when I call you later, bring three G's to that address."

In perfect accordance with the plan, Outlaw found several days later, upon returning from chow, a shoebox-size package tightly wrapped in brown paper under his pillow. His heart jumped. Making sure no officers were nearby, he unwrapped the cardboard box and marveled at its contents: a revolver. A thing of wondrous, sheer-black metallic beauty and power. He looked at it for some time and smiled. This was a game changer. But Outlaw didn't want the gun in his cell. Within two hours, he'd bundled it up in a towel, slipped the towel into his briefs, and passed the whole thing off to one of his cadre. Outlaw didn't even know where exactly the man stored it, but he trusted his associate to hide it in some remote corner where the guards would not even know to look. Outlaw knew it was available to him. That was the important thing.

It all came to the fore in the yard on a hot, dusty July day. Hundreds of prisoners were milling around, and armed guards were in the gun towers. Mike Lajoie was in the yard, and witnessed an escalating tension between the Latin Kings and Tiny Piskorski's crew. Milan and Millett had twenty of their men, and Tiny fifteen of his. The two groups approached each other. Piskorski and the two Latin King leaders stared each other down. Lajoie was twenty yards away and it felt to him like a riot was about to happen. He feared for his safety. The staff were totally undermanned, five inmates to each guard.

Lajoie saw Outlaw walk up with two of his guys and huddle with the Latin King leaders and Tiny. To Lajoie, this was the moment it could blow. But then everything stopped. The Latin Kings walked away, and the white prisoners did too, to separate areas in the yard, like boxers going back to their neutral corners.

It was not until years later that Lajoie found out what happened in the yard that day. A member of the staff had found a gun hidden in the library. The pages of a book had been carved out to contain it. Lajoie never found out whom the gun belonged to, although he always suspected Outlaw. Then, many years after the discovery of the gun, Lajoie was conducting an exit interview with one of Outlaw's lieutenants just before his release. "I have to ask you, and I won't hold this against you: Did Outlaw have a gun that day out in the yard?" "Yes, he did," the inmate said. "He flashed it at Milan and Millett and Piskorski. You wouldn't believe the shock on their faces." Outlaw brought a gun to the proverbial knife fight, and the other shot callers had no choice but to retreat. Outlaw had become the king of the shot callers.

Lajoie was starting to think of Outlaw as the Biggie Smalls of Somers, a once-in-a-generation prisoner with a unique combination of smarts, size, and charm. Possibly, even, some kind of genius.

BUT AWAY FROM his public image, beyond all the bravado and the swagger and his relentless displays of guile, Outlaw was exhausted and scared. Every night, after his cell door clanged shut, he felt relieved and oddly free. No one could bother him for the next precious seven hours. He had no need to live up to the image that he'd created, the larger-than-life intimidator. He could be himself: a frightened man embarrassed to his marrow about the swamp his life had become. Tainted by one misdeed after another, Outlaw's family was starting to write him off. He still retained some pride about the Jungle Boys, but increasingly, the entire enterprise felt like an endless, blood-dripping nightmare. He'd done horrible things and all the ego in the world couldn't undo that. The damage was irrevocable. But at the same time he couldn't let down his defenses for a second: when the cell doors popped open at 5:30 A.M., as they did every day, they—the guards, the Latin Kings, Tiny, whoever—would be waiting. The mask had to go back on.

Outlaw didn't really sleep for years, perhaps one to three hours a night. He embarked on an endless conversation with himself during those seemingly infinite, wakeful night hours. He would toss and turn from the constant churn in his mind: the ceaseless and agonizing what-ifs. What if he'd let someone else shoot Sterling Williams? What if he'd let Fly Boy's insults just go? What if he'd taken the Jungle Boys' money and gone to North Carolina? What if he'd gotten that summer job in the parks? What if he'd never said yes to Ducky and Butchy in the first place?

While he enjoyed seeing Camile during her frequent visits, he was depressed that Pearl didn't come to see him often. He knew that the distance from New Haven was the main reason for her absence, along with how little time she had between her two jobs, but he also understood that there was more to her diminished presence than mere logistics. She was ashamed of him. And angry too—angry at her son who had debased her. Outlaw appreciated the occasional visits with his children, which had now grown to five offspring, but he wasn't always on the best terms with their mothers. The kids were now two to four years old, and as they aged, Outlaw started to feel more self-conscious and humiliated in their presence.

And somewhere in the middle of it all, his grandfather, his former fellow Somers inmate, died in New Haven. Outlaw was brought by the state police to the funeral at a Baptist church in New Haven. He was not allowed to attend the service itself but was permitted to view the body beforehand. He had no idea what his grandfather looked like; they had never met. He was just about to look at his grandfather's face in repose, but at the last moment turned his head away and walked out. *Why should I even bother to look at him?* His grandfather had done nothing for him, nor had his son, Outlaw's father. He didn't even owe his grandfather as much as a glance. Outlaw asked the state police to take him back to Somers. The only good thing about the trip was that he'd been able to steal a couple of words with the Jungle Boys on the sidewalk.

Outlaw had one more unhappy visit back to New Haven during his first year of prison. Fly Boy had reappeared in town and was now going to press charges for the shooting on June 29, 1988. A trial was hastily put together in New Haven Superior Court for March 23, 1990. Outlaw was driven in handcuffs to the New Haven courthouse, where he met Ira Grudberg who told him once again not to take the stand. Outlaw tried to direct smoldering looks toward Fly Boy during the trial, but he no longer had the ability to personally back up any threats of intimidation, and he was hearing that the Jungle Boys were in increasing disarray without his presence. Given that Outlaw was serving sixty years for murder and another assault, he had absolutely no credibility with the jury, and even Ira Grudberg couldn't work his magic. Twenty-five more years were added to Outlaw's sentence, changing his date of release to April 7, 2073, at which point Outlaw would be 105 years old.

At this point, Outlaw was less and less concerned about the course of his life. The realization was clear: he would surely die in prison now.

IN THE MONTHS that followed the expanded sentence, Grudberg decided to work on an appeal to Outlaw's first conviction based on the fact that Fitzroy Phillip never appeared and never testified at the trial. Grudberg had always complained that he had never had a chance to cross-examine Phillip, and that the transcripts from the probable cause hearing that Phillip participated in a few weeks after the shooting were not valid in and of themselves and should not have been admitted to trial. After months of preparation, including a couple of last-minute visits with Outlaw to apprise him of the strategy, Grudberg brought his appeal to the Connecticut Supreme Court on September 28, 1990.

Outlaw didn't expect much to come of Grudberg's campaign. He was realistic about the high bar for such appeals, and Grudberg had already failed him twice in court—not that Outlaw particularly blamed him for it. At this point Outlaw felt the best chance for evading his eighty-five-year sentence was to escape from prison.

In fact, Outlaw had pretty much forgotten about the appeal altogether when two months later, on November 20, 1990, he was lying in his cell bed at just after twelve noon. Outlaw's fellow inmate Richard Crafts, who was known as "the Wood Chipper" and whose cell adjoined Outlaw's, excitedly called out to him. Crafts, a former airline pilot, was so called because he'd murdered his wife and shredded her remains in a chipper.

"Turn on your TV! You're going to be on TV!" Crafts shouted to Outlaw. Crafts had been watching the twelve o'clock news on WFSB, the local CBS affiliate in Hartford. Right before a commercial break, the newscaster had said in the breathy, seductive manner of local television journalists, "Up next, breaking news about William Outlaw, reputed New Haven drug kingpin."

Outlaw switched on the television at the foot of his bed and agonizingly sat through a commercial until the broadcast returned. From her anchor's desk, Gayle King, then a young newscaster, announced that the Supreme Court had decided in Outlaw's favor. They agreed with Grudberg's argument that Phillip's testimony was not valid, and that charges for shooting Phillip and murdering Williams should be dropped, and a new trial ordered. Potentially sixty years could be taken off his sentence.

Outlaw dropped his head in silence. He felt his body sink into the mattress. It took some seconds for any kind of coherent thought to form in his brain. Sixty years of his life had potentially been delivered back to him in a moment's notice. He realized the unthinkable: his release was no longer 2073, but possibly 2013. A God-sent miracle. It could be no other thing. Tears rolled down his face. His thoughts went back to his prayer on that first night at Somers, when he lay on the floor: *God, if I ever get out of here, I will do something with my life—I don't know what, but I will do something . . . something better.* If things worked out, he was going to have to make good on that promise.

These reveries were broken by the cheers of inmates. He heard his name chanted from cells up and down the unit. It took Outlaw a moment to catch up to what might have happened—was *any* of this real?—and then he realized that many of his block mates had been watching the very same news.

Moments later, it was mealtime. The guards popped the cell doors open.

"Yo, J, congratulations!"

"You're gonna be a free man. Can you take me with you!"

High fives and embraces all around.

At lunch, where he couldn't eat as the joyous numb ecstasy continued, he felt his swagger returning. Sixty years! Sixty years of his life could be returned to him on a golden platter.

The decision was covered the next day in the *New York Times*:

November 21, 1990

NEW TRIAL ORDERED FOR MURDER SUSPECT

The Connecticut Supreme Court has thrown out a murder conviction and ordered a new trial for William Outlaw, whom the police have accused of

running a violent drug ring in New Haven. The court ruled that the jury
should not have heard transcripts of earlier testimony from a witness who
failed to appear for the actual trial.

Mr. Outlaw was sentenced to 60 years in prison in July 1989 for the
shooting death of Sterling Williams in September 1988. The police said Mr.
Outlaw, 23 years old, killed Mr. Williams in a dispute over drug dealing.

Mr. Outlaw is also serving 25 years for second-degree assault in another
shooting, in June 1988. Ira Grudberg, Mr. Outlaw's lawyer, had argued
from the start that the testimony by the witness, Fitzroy Phillip, was unfair
because he did not have an opportunity to cross-examine Mr. Phillip.

Mr. Phillip testified at a hearing in October 1988 that Mr. Outlaw was
the man who wounded him in the back and leg and killed Mr. Williams. But
by the time the trial started Mr. Phillip had disappeared.

Still walking on air, Outlaw was visited a few days later by Tom Ullman, a court-appointed public defender from New Haven in his early
forties. He explained he was Outlaw's new attorney, that the cash that
Outlaw had paid Grudberg had run out and Grudberg had written a letter to the public defender's office saying he would no longer represent
him. Outlaw was disappointed by the news but took an immediate liking
to Ullman, who had an appealing directness.

"You're from New York?" Outlaw said. He detected a New York
accent.

"Yeah, Queens," Ullmann said. Outlaw liked that. New York meant
toughness.

Ullman met with Outlaw a number of times. He exuded a passion
for the law, so much so that he smiled broadly even when talking about
the most arcane legal matters. Ullman advised Outlaw to try and work
out a deal rather than be retried for the Williams and Phillip cases. If
another trial occurred, Outlaw's sixty years could easily be restored.
Ullman suggested that Outlaw, in exchange for not pursuing another
trial, accept a downgrading of the Williams charge to manslaughter,
which carried 20 years. Outlaw would then be able to serve the Williams
sentence concurrently with time served on the Fly Boy shooting, leaving

Outlaw, finally, with a newly combined sentence of thirty years altogether. Outlaw agreed to the strategy, having implicit faith in Ullman's
authenticity and professionalism. Ullman returned a few weeks later to
say that the deal had been accepted. His new sentence was thirty years,
and with good behavior in prison that might be downgraded over time
to twenty or so years in prison. Outlaw hugged Ullman and realized
he'd lucked into being represented by one of the best public defenders
in the country.

GRATEFUL AS HE was for the miraculous reprieve, Outlaw was not yet
ready to give up his shot-calling ways in prison. He was still going to be
locked up for decades and needed to maintain his status. Any diminu-
tion of his brutish ways would make him vulnerable—there were plenty
of up-and-coming inmates who would go after him in a second if he let
down his guard.

What Outlaw did not know was that months before, fed up with his
antics and those of other shot callers, the Department of Correction had
formed an anti-gang task force led by Mike Lajoie and Brian Murphy, a
senior corrections officer and member of the Northern Boys, in an attempt
to retake control of their prisons. Lajoie and his colleagues were adamant
in their determination to restore order. They dropped microphones sur-
reptitiously into cells; cut off all family visits, phone calls, and privileges
for suspected gang members; and created a special gang-members-only
block of the prison, even putting rival gang members together in the same
cell as a kind of experiment to see what would happen. When he was
ridiculed for this tactic, Lajoie merely shrugged. He found that former
adversaries would often bond together and cancel their aggression out.

But all the task force members agreed that the first step of the initiative was to remove the "heavy hitters" from the system and transfer them
to out-of-state or federal prisons to break their sphere of influence. Or
as the plan's principal architect Brian Murphy, who went on to become
the department's commissioner two decades later, said, "We had to cut
the snake off at the head." The task force had begun the process a year

earlier, by sending Sonny, a leader of the Latin Kings, to Lewisburg, the notorious federal prison in Pennsylvania. Six months later word had come back that he was dead, stabbed to death by another inmate.

But Outlaw was oblivious to Lajoie's plans. For a time, in the aftermath of his legal success, things were pleasantly normal, at least for prison. Outlaw came up with the outrageous idea of getting married to Camile. Camile was agreeable, strangely enthusiastic, given the circumstances of the planned betrothal. Despite it all, Camile loved Outlaw, and as devout as she was, it seemed like the right thing to do, the Christian thing to do, not to forsake Outlaw in his hour of need. And Outlaw, despite his infidelities, loved Camile. With the warden's permission, weddings were allowed in prison, in the old courtroom chambers—a medium-size room on the ground floor, the walls painted a drab institutional beige. Camile and Outlaw were married on March 5, 1992, at 11 A.M. by the prison chaplain. Camile wore a blue dress and Outlaw his prison uniform, which he had ironed for the occasion; he had arranged, through old Jungle Boys friends, to get her a beautiful gold ring with three diamonds. Camile felt blissful. She had done the right thing.

Simultaneously, Mike Lajoie and his team were actively trying to rid Somers of Outlaw. The problem was, they had to catch him doing something illegal. The authorities needed a provable justification for their actions; until they had specific evidence, they were powerless to transfer him. On April 24, 1992, they got what they needed. As reported later in the *Hartford Courant*, a telephone wiretap on that date captured Outlaw talking to one of the Jungle Boys. The FBI, the New Haven police, Lajoie and other corrections officials had conspired to monitor his outgoing calls from prison. "Hey, yo," Outlaw said to his associate in the recording. "When you talk to Shark, say JB said to run him two bundles of weed." A bundle is about twenty-five bags of marijuana. Outlaw was caught in the act of ordering drugs to be brought into prison.

Even more damning was another recorded conversation that was not reported in the newspaper. One of the Jungle Boys called Outlaw to report a shoot-out in the Jungle between the gang and the police. Outlaw was recorded saying, "You guys fucked up. You weren't aggressive enough. You had an opportunity to shoot the fucking police. If I'd been there, I would have fucking killed one of those pigs."

The Jungle Boys were now largely run by Rodney, who lived in the Jungle, making it far easier for law enforcement to track the gang's activities. On top of that, New Haven police were able to take advantage of the newly enforced RICO (Racketeer Influenced and Corrupt Organizations) Act, created in 1970 to target the mob and other gangs. Under RICO provisions, a vast and unprecedented multiagency task force of local, state, and federal officials targeted the Jungle Boys as the first step in bringing New Haven back to safety. They observed the gang's operations for months, including using surveillance posts on top of the train station.

On June 22, 1992, they descended. As reported in the *New York Times:*

> A team of 150 law enforcement officials, which included agents from
> the Bureau of Alcohol, Tobacco and Firearms, the Federal Bureau of
> Investigation, the Drug Enforcement Administration along with state
> and local police, raided a housing project in New Haven and arrested
> 14 members of the Jungle Boys and charged them with conspiring to sell
> cocaine and violating Federal firearms regulations. During the raid, agents
> confiscated six firearms, including a sawed-off shotgun and a 12-gauge
> pump shotgun. The investigation leading to the arrest began in February
> and included undercover purchases of narcotics and electronics surveillance.
> If convicted of the conspiracy charges, each defendant faces a minimum of 5
> to 40 years in prison.

The day after one of the largest busts in New Haven history, the FBI and the New Haven police held a joint press conference inside the walled confines of the Jungle. Fifteen politicians, police officers, and officials appeared before the television cameras, standing in front of Jungle Boys graffiti. They announced triumphantly that a scourge had been eliminated from the city. The police chief said that half of the murders in New Haven were attributable to gangs like the Jungle Boys. "On a per-capita basis, if this were New York or Boston, you'd have the National Guard in here," said the chief of the Drug Enforcement Agency's Connecticut operations. The officials declared that the police department would open a substation within the Jungle, and that law enforcement agencies and Yale architecture students would work to reconfigure parts of the Jungle to make it harder to deal drugs out in the open.

The evidence used to convict the Jungle Boys included the testimony of twenty-four witnesses; seventy-five tape-recorded intercepted telephone conversations; narcotics purchased by undercover agents; narcotics, narcotics paraphernalia, and firearms of the gang, seized by the government; and the video and photographic surveillance. Numerous Jungle Boys were sentenced to between 168 months and 328 months in prison.

It was the end of the Jungle Boys. Hearing about the massacre of his gang, Outlaw was furious. He blamed the others, and said this would not have happened had he remained in charge. But at the same time, he thought, he may have dodged another extraordinary bullet. Had he not already been in prison, he could have received at least what Rodney did. He had lucked out again. It was as if he had been reborn—not once, but twice.

Two weeks later, a guard brought Outlaw to a meeting room, where waiting for him were the deputy warden, the former beat cop and New Haven police detective Ed Kendall, who had first attempted to counsel Outlaw when he was a freshman at Wilbur Cross, and a man who introduced himself as an FBI agent. The agent smiled warmly at Outlaw, extending his hand. Outlaw didn't shake it. The agent told Outlaw to have a seat, but he chose to stand.

"We have all kinds of evidence against you," the agent said. Outlaw said nothing. "I hear you're a smart man," the FBI agent continued in a friendly, almost collegial manner. "If you tell me everything about the Jungle Boys we've arrested in New Haven, we can get you out of prison a lot earlier. We'll cop you a deal."

They had papers for Outlaw to sign, some kind of a plea bargain, and asked him to examine them. "Fuck you," Outlaw said. Outlaw wasn't upset at the agent—to Outlaw, he was just doing his job—but rather at

Kendall. "You faggot ass cop," Outlaw screamed. "You belittle me by showing up and asking me to do this. Get the fuck out of here!"

The deputy warden told him to calm down.

"Calm down my motherfucking ass, you got some motherfucking nerve to come in and ask me to sign. Don't you know me better than that?" he said, now up in Kendall's face.

The FBI agent's tone changed completely. "You're gonna get indicted, you know," he said sharply. "We have everything that we need."

It was anathema for Outlaw to cooperate. It would destroy everything he had created. "Do what you need to do," Outlaw replied. "I don't cooperate with cops and Feds. And, Kendall, you of all people should fucking know that!" Outlaw couldn't believe Kendall's ignorance of his character. How could Kendall not know he would never be a snitch?

SOME WEEKS LATER, on September 24, 1992, four years to the day after he had killed Sterling Williams, twenty-four-year-old Outlaw was awakened in his prison cell by two guards at three in the morning. The first officer shone a flashlight in Outlaw's face. Outlaw, who had not been sleeping soundly, rose from bed. In his usual flaunting of the rules, he was wearing street clothes, a silky maroon sweat suit and Nike sneakers. The second officer said, "Outlaw, you have a visitor. Put on the uniform of the day." The "uniform of the day" meant khaki pants, white T-shirt, and black leather boots. The hour of the guard's request was unusual, but Outlaw was not particularly concerned. What could the system possibly do to him that it hadn't already done? The two officers led Outlaw down the dark hallway. They could hear only the intermittent snoring of prisoners and indistinct rustlings of the few who were up at that time of the night.

"Where you going, Juneboy?" came a voice from one of the cells. "What's going on?"

"Got a visitor," Outlaw said.

"This time of night?" said the voice from the darkness.

The trio walked past a gate and a hall keeper's desk with a guard stationed at it. As they approached the entrance to the visitors room, Outlaw slowed up. But one of the guards said, "Keep going."

This struck Outlaw as odd, and he felt, for the first time, a ripple of anxiety. The guards led him another hundred feet down the hallway. They approached another door, a handsome dark wooden door. This was the court chambers, where he and Camile had been married. "Go in," the guard said.

Inside Outlaw found a group of eight men waiting for him, including Mike Lajoie, three guards, the deputy warden, and three other men—lean, fit military types, each wearing navy-blue jackets and pants and military-style boots. Written on their jackets in yellow letters was US MARSHAL. On the floor in front of them was an orange jumpsuit, size XXXL. Next to the jumpsuit were iron leg shackles, a waist chain, and handcuffs.

"Mr. Outlaw, you are being transferred out of the Connecticut Department of Correction as a result of the Interstate Compact," one of the marshals said. "A box of your belongings will follow you to your final destination." Outlaw immediately thought of Sonny, killed not long after leaving Somers and being transferred to the Feds.

Outlaw knew what the Interstate Compact was. He wasn't aware of all the legalities, but he knew that it was a mechanism to transfer prisoners that a state prison could no longer manage to either a state prison in a different part of the country, or to federal prison, known to be much tougher than state prison. Outlaw was aware that the Interstate Compact was implemented only in extremely rare situations, and he never believed that the authorities would actually dare apply it to him.

A marshal instructed Outlaw to strip to his underwear, and examined him from head to toe. He first told Outlaw to open his mouth, and looked inside. Outlaw was then told to turn around and spread his buttocks, squat down, and cough. This procedure was done in lieu of a digital exam of the anus. The Connecticut prison officials observed the proceedings. Outlaw looked across at them defiantly and thought that they were smiling, inwardly celebrating that their nemesis was now getting his due. *Assholes, all of them*, Outlaw thought—*except for Lajoie*. Lajoie, even given what he was doing to him now, was still all right to Outlaw—simply doing his job honestly. No hard feelings between them.

The marshals then told Outlaw to put on the jumpsuit and placed manacles around his ankles connected by a short chain, and locked them. Then they put him in handcuffs and ran a heavy steel chain around Outlaw's waist, connected to the handcuffs. When the process was done, Outlaw could move only enough to shuffle, and barely at that.

The lead marshal grabbed Outlaw's Buddha chain and threw it to the ground. "You won't be needing that where you're going," he said.

The same marshal spotted his wedding ring, unclipped a ring cutter from his belt, wedged the base of the tool, a thin strip of metal, under the ring, and activated its mini-saw. The ring broke apart with a popping sound.

"You're going to have to pay for that," Outlaw said.

"Don't worry. We will," the marshal said.

Outlaw started laughing. The marshals were unused to offenders laughing at them, particularly one that was outnumbered eight to one. They asked what he could possibly find amusing at that moment.

Outlaw sneered. "Because you waited for me to get into chains before you dared cut that ring off my finger," he replied.

The business done, Outlaw was led out of the courtroom and down the hallway to the main entrance of Somers. Outside the prison was a black Lincoln Continental. The marshals told Outlaw to get in the backseat.

"See you later, Outlaw," said Lajoie. Indeed, they would see each other again, years in the future.

In the back of the car, Outlaw stared into space. He couldn't show it but he was stunned. *They nailed me.* He was outnumbered and out-manned. Nobody could help him now. He thought of Pearl, and his wife, and his children. The marshals got into the car and drove Outlaw off into the night.

THE OBSCURE FEDERAL law by which Outlaw was being transferred out of Connecticut—the Interstate Compact—dated back to 1936, when Congress allowed states to "enter into a compact with any of the United States for the mutual helpfulness in relation to persons convicted of crimes or offenses." What this meant in practice was that an individual state could, in effect, transfer its most notorious prisoner and in return receive another state's worst prisoner. The hope was that a previously incorrigible felon, away from his home turf, would no longer enjoy his sphere of influence—often developed over the course of decades—over his peers, and a negative force in a given system would immediately be excised.

But with the deployment of the Interstate Compact in Outlaw's case, there was no prisoner exchange. For the privilege of banishing Outlaw, the State of Connecticut was willing to pay the Federal Bureau of Prisons $30,000 a year for his housing and supervision, and receive nothing in return, other than the peace of mind that William Juneboy Outlaw III was no longer within the state's borders. He was effectively being banished from Connecticut.

Prison insiders had another, less formal way of describing the phenomenon: "bus therapy."

CHAPTER 5

Bus Therapy

Federal prison, Lewisburg, Pennsylvania.
National Archives.

AFTER EXITING THE GROUNDS OF SOMERS PRISON, THE LINCOLN CONtinental drove south to Hartford and then west to Danbury, close to the New York border. The marshals pulled into the compounds of the Danbury Federal Correctional Institution, mainly a women's facility. For one hour, Outlaw and the three marshals sat in the car in the parking lot. Outlaw said nothing while the marshals chatted idly about sports. Finally, a Greyhound-style bus, painted white with an American flag on its side, pulled up. Outlaw watched as prisoners shuffled off the bus, in the same arm and leg shackles that he wore. The marshals instructed Outlaw to get out of the car and directed him to the processing room inside the prison, where he was inspected bodily, photographed and fingerprinted, and then led back to the bus. Inside it was like no bus that he had ever seen. It was fitted with hard slate-colored benches, and behind the third row of benches was a floor-to-ceiling mesh screen with a door, creating, in effect, a cage for the prisoners inside the bus. Outlaw was instructed to go to the last bench next to a marshal who sat with a rifle on his knee. Over the next hour, ten additional prisoners filed in, most of them African American, older men with hard faces and dead-looking eyes. By contrast, Outlaw with his smooth and open face looked like a teenager, despite his girth. The bus, manned by four marshals in all, rolled out of the complex. None of the passengers had any idea where they were going.

Later that afternoon, after six hours of uninterrupted travel, Outlaw found himself in the middle of Pennsylvania. He looked out the window at the rolling fields, which were much more open than those in Connecticut. Out of nowhere, a huge brick building—like some sort of grim castle—appeared on the horizon. The sprawling structure was a quarter mile wide, surrounded by a thirty-foot-high concrete wall, with redbrick watchtowers at each corner. Presiding over all of it was a two-hundred-foot-tall Italian Renaissance–style tower, looking like something out of medieval Florence. One of the marshals mumbled that they were soon to arrive at the Lewisburg Federal Penitentiary. Outlaw knew a little about Lewisburg: it was one of the most infamous and dangerous of the federal prisons, where Whitey Bulger, Al Capone, and John Gotti had served time. He also knew it was the place that Sonny had just been stabbed to death. Outlaw shuddered.

WHEN LEWISBURG FEDERAL Penitentiary first opened in 1932, its creators hoped that the facility, which with its towers and arches and battlements resembled a monastery more than a prison, would have the effect of uplifting the souls of its inhabitants. The site had been selected carefully, on a tract of land the size of New York's Central Park in the wilds of Pennsylvania. Lewisburg's architect was attracted by the "beneficial and stimulating qualities of the sunshine and fresh air [that were] more readily available in such settings." Indeed, early on, inmates raised poultry and cattle, and cultivated hay, corn, clover, soybeans, alfalfa, sorghum, and potatoes. They played baseball and basketball, boxed, and lifted weights as well as performed in theatrical groups. Famous performers occasionally passed through—Louis Armstrong and his band played there in the 1950s.

But the Lewisburg at which Outlaw arrived on September 24, 1992, bore no resemblance to the early days of the institution. Beset by decades of rising prisoner populations, race riots, guard brutality, and budget woes, the once pristine campus was in disrepair. The Gothic nature of the facility, once meant to be inspirational, had taken on a decrepit, disturbing feeling, and the lofty principles upon which the institution had been founded had long since been abandoned. By the 1960s and 1970s, amid rising crime and high rates of criminal recidivism (nationally, about half of inmates released from prison were incarcerated again within three years), the Bureau of Prisons largely abandoned any interest in rehabilitation, replacing that ideal with the liberal use of solitary confinement and lockdowns.

Outlaw was called off the bus. As he surveyed the behemoth of a building in front of him, almost too big to fathom, he saw that everything about Lewisburg prison was red. Red brick, red roofs, and—as he would soon learn from its residents—red floors from the blood that was regularly spilled on them. As he attempted to take the view in, he was intimidated and overwhelmed, but at the same time, he felt a curious sense of excitement—even pride—surge within him. Getting to Lewisburg meant, in a certain way, that he had entered the big time. He thought, *When I became a gangster, I wanted to be the best. The biggest and the baddest. I only heard about places like Lewisburg in books and rap songs. But now I am actually here.* In his mind, he had transcended the local level and entered the rarefied status of a nationally recognized gangster. Even in the uncertainty and danger, something about that made Outlaw feel good.

He went through a twenty-foot-tall steel gate. The guards took his shackles off in a processing area, and Outlaw rubbed his liberated hands and wrists. He was told to remove his jumpsuit and issued a flimsy dark blue Lewisburg prison uniform. The suit was made of paper, not cloth, to prevent prisoners from fashioning ropes out of their garments. Outlaw was then escorted by a guard on a long walk through A Block, then B, all the way to F. Along the way, men shouted out from their cells. "Where y'all come from?" "What up, big man?" "See you on the compound!" Outlaw did not know much about federal prisons, but he knew enough not to respond. He was led down two flights of steps into a basement. "You've been assigned to solitary confinement," the guard said nonchalantly, leading Outlaw into a bone-cold, damp, and poorly lit hallway. Outlaw was shoved into a five-by-nine-foot cell. Inside was a horizontal plate of steel bolted into the wall, on which sat a foam mattress, and a metal commode and a sink, with no mirror. The faucets dripped. The walls were made out of the omnipresent red brick, the grout failing. The metal door had a narrow horizontal slot in the middle through which meals would be brought three times a day. The place smelled of feces; some of the more disturbed prisoners were known to cake the walls with their own shit. In his three years at Somers, Outlaw had never seen anything like this. This was a dungeon.

The guard shut the door behind him with a familiar metal clang. Stunned by the day's events, Outlaw lay on the bed and wrapped himself in a blanket. Suddenly it was just him in this tiny cell, 250 miles from home. He thought he might rest, but, suddenly, he heard inmates calling out to him: "Yo, newbie, where you from? Who the fuck are you?" *How had they seen him come in? Where were the voices coming from?* Startled, he looked up and saw heating vents, which seemed to connect all the rooms. The voices of inmates from other cells were echoing through them, fun-house style.

He began to sob. If only the Jungle Boys could see him now.

BACK IN CONNECTICUT, Camile heard that Outlaw had been abruptly transferred, but she had no idea where he had been taken. Nobody at Somers did. Camile drove to the prison and demanded that the warden tell her where Outlaw had been sent, but the warden refused to meet with her, and his staff said they were not free to disclose any information. Camile stressed to the indifferent staff that she recently had become his wife, but to no avail. She then went to the Bureau of Prisons office in Hartford, where she was told again that they could not release the information. She held out her marriage certificate. "Okay," a clerk said, giving in. "I shouldn't tell you this, but he's in Pennsylvania." She took the next day off from work and drove to Lewisburg.

"How did you find me?" Outlaw said, in the dismal waiting room.

"That's what I do," she said. "I solve problems."

It had only been a week since he'd left Connecticut, but Camile arrived to find Outlaw a broken man. He was weeping. He said it was so loud in solitary that he couldn't sleep. His sense of humor, his constant irreverence, his ebullience, were long gone. He had lost weight. Camile couldn't fathom how quickly his descent had occurred.

"What can I do for you?"

"Nothing," he said listlessly. "What's there to do?" He wouldn't look her in the eye.

So she just sat with him. She hugged him when the guards weren't looking. He barely responded.

"Thank you for coming," he said when she left.

Back in Connecticut, Mike Lajoie and the Northern Boys were thrilled by Outlaw's departure. There had been an immediate downtick in disruptive behavior starting the day he had left.

IN THE DUNGEON, Outlaw tried to keep his mind occupied by doing mental exercises and push-ups and sit-ups and jumping jacks. As much as he could, Outlaw listened to the guards talk outside his cell, trying to pick up any information he could about Lewisburg, and the federal system generally. After a few days, he figured out a way to play chess with inmates across the way, even though, of course, they couldn't see each other. Each inmate scratched a mini-chessboard onto a bar of soap and called out the moves through the vents. The game at least brought him some distraction. Breakfast came through the slot at the ungodly hour of

3 A.M.: a piece of toast, an egg, an expired pint of milk. At five-thirty the guards entered, made him strip to check that he had no contraband—not that he could have possibly smuggled anything into his cell—and examined him from head to toe, including an examination of his open mouth and the usual squat and cough, then cuffed him up and put him in the shower or the rec yard for an hour. At nine A.M. came lunch: bullet peas, mystery meat, pudding. Three P.M. brought dinner. This was the final human contact of the day: that is, if you could define a tray sliding through a door as human contact. After that, Outlaw tried to sleep, which proved to be a fitful undertaking. Each day the cell seemed to get a little smaller.

As it turned out, Outlaw was at Lewisburg for a short time. At 3 A.M. on October 6, 1992, in a procedure eerily identical to what had transpired at Somers two weeks earlier, he was woken by marshals, put in shackles, and driven to the Harrisburg airport. Along with a dozen other inmates, he was placed on a white 747 operated by the U.S. Marshals Service, often referred to as "Con Air" or "the Midnight Express." Outlaw was flown halfway around the country, as the plane made stops in Cincinnati, St. Louis, and Oklahoma City. That night he was placed in solitary in El Reno, a federal prison in Oklahoma. The next morning he was put on another plane and flown to San Francisco, then Seattle, then Denver, before finally touching down at Kansas City International Airport, which, insanely, was just three hundred miles from where Outlaw had awakened that morning. Cadres of marshals in body armor and with automatic assault rifles awaited him on the tarmac of the airport. A marshal said to him in a Midwestern twang, "Welcome to Kansas. The only thing that we have that's your color out here are queers and steers."

Soon enough, Outlaw found himself in an armored van with other prisoners. Looking out the window as the van exited the airport, Outlaw saw the suburban neighborhoods emptying out, replaced by prairie and honey-colored autumnal fields of wheat. Solitary hawks wheeled in the blue and limitless sky, which in its infinitude was unlike any sky he had seen in Connecticut. The autumn heat rose off the highway.

Forty-five minutes later, Outlaw spotted a prison complex surrounded by fifty-foot-tall barbed-wire fences. At its center was a soaring central building with a white cupola resembling the United States Capitol that rose hundreds of feet above the Kansan plains. A sign said: UNITED STATES PENITENTIARY LEAVENWORTH. Once he was inside, the intake process, including medical and mental health screening, and orientation to the facility, was more professional and seamless than anything Outlaw had previously undergone. He was notified of the system of inmate counts, which would occur at 12:01 A.M., 3:00 A.M., 4:45 A.M., 4:00 P.M., and 9:30 P.M. At these times he would need to stand in his cell as staff passed through the units to ensure that all prisoners were accounted for.

A famous saying within the Bureau of Prisons is that "no prisoner is sent directly to Leavenworth. He has to earn his way there by fucking up everywhere else." Indeed, Outlaw—by having shown absolutely no willingness to cooperate with the authorities at any point in his prison odyssey, by being convicted of murder, by being a known shot caller in Connecticut, by being caught in the act of arranging to bring drugs into prison, by not cooperating with the FBI in the Jungle Boys investigation—had been deemed by the system to have "fucked up everywhere else." Accordingly, he had been sent to the first and most notorious institution in the Bureau of Prisons, often called "the Harvard of federal prisons." When Outlaw arrived in Leavenworth, there were 1.3 million incarcerated individuals in the country, the vast majority of whom were housed in state prisons. Only seventy thousand inmates—generally considered significantly more dangerous and callous than the state prisoners—were relegated to the federal system. Within the federal system, there were five tiers of facilities: Level One being the least secure and Level Five being the most. Of Level Five prisons, there were only two in the country—Marion, in Illinois, and Leavenworth—and between those two, Leavenworth was considered the most formidable. Former residents included James Earl Ray (the assassin of Martin Luther King), Whitey Bulger, Machine Gun Kelly, and Perry Edward Smith and Richard Hickock, the killers of the Clutter family depicted in Truman Capote's *In Cold Blood*, who were also electrocuted there. A fundamental difference between Leavenworth and state prisons like Somers was that most residents at Leavenworth were sentenced to thirty years or more, leaving them little incentive for compliance or self-improvement. Indeed, three months before Outlaw arrived at Leavenworth, a riot involving three hundred prisoners broke out during a screening of *The Silence of the Lambs*—perhaps not the wisest choice of entertainment for inmates. The prisoners took over the auditorium and the rec yard, and one inmate was killed. The siege lasted seven hours before being brought under control by the National Guard.

In creating his own path to Leavenworth, Outlaw now found himself among the sixteen hundred most unrepentant men in the country.

FROM THE MOMENT he arrived in Kansas, Outlaw was a stranger in a strange land. Everything was different. Despite the physical order and precision of the facility, the mood felt incendiary, as if it could burst into flames at any time. Outlaw saw that his peers were notably more grizzled and hostile than the average inmate at Somers, and a much higher percentage of prisoners—about half—were white. Most of them too were more physically imposing and muscular—the accumulation of decades of weight-lifting, an additional and necessary armor to withstand the federal system. In the yard, even in October, the sun was hotter and more piercing than in the Northeast. Even inside the prison, the air was arid.

The whole place felt like a time warp to the 1950s. People moved more languidly than back east, and there was hardly a black man on the staff. Everything inside—the cinder-block walls, the tile floors, even the doors and bars of the cells—was a gleaming, almost alabaster white. To add to the dissonance, Outlaw found that from his arrival other inmates kept on asking him how things were in Philadelphia. Outlaw was confused until someone told him that his inmate number, which had been assigned at Lewisburg, had digits that reflected that he was a Pennsylvania resident. (Each state has inmate number codes, which any veteran prisoner knows by heart.) Upon reflection, Outlaw didn't particularly mind this error: Philly surely had more street cred than Connecticut, and maybe that would help.

This confusion underscored the fact that nobody at Leavenworth had any idea who he was, which was no doubt the intention of the Connecticut authorities: to cut his sphere of influence out from under him and make him vulnerable, even perhaps testing to see if he could survive. Indeed, to his panic, Outlaw discovered shortly after arriving in Kansas that he had been sent from Connecticut without his "paperwork": an inmate's key identifying information, which lists their charges and movement throughout the correctional system. Such documents serve as a kind of passport as prisoners go through the system, providing proof of identity and, even more important, one's credibility. Having certain crimes, like murders and armed robberies, on one's record is advantageous, while others—sex crimes, crimes against children—can get an inmate killed. Without his paperwork, Outlaw would be suspected of being a child molester or a snitch.

On Outlaw's first night at Leavenworth he had finally, blessedly, arrived at the privacy of his cell, high on the block on the fifth and top tier, secure at least for the next ten hours, until he would be taken out to the yard at 5:30 A.M. He was washing his face in the sink in his cell when he heard, "Hey, you! Nigger!"

Outlaw swiveled around to see four white men outside his cell. They had shaved heads, handlebar mustaches, and tattoos of Nazi crosses and four-leaf clovers on their forearms. One of the men leaned in close to the bars of the cell and said, "Are you a snitch?"

"No," Outlaw said. "I'm not a motherfucking snitch."

One of the men—who had battle-tested, penetrating blue eyes—said, "Well, we'll find out about that. If you are, we'll come back and kill you, throw you over the tier."

Outlaw kept his face completely impassive, like a boxer before the opening bell of a fight, as the men threatened him from across the bars. He couldn't afford to blink or twitch. Inwardly, though, he noted that he'd only been on the unit for half an hour and his life had already been threatened. He had only one thought: *I had better get a knife as soon as possible, because they are going to come back and kill me.*

Having delivered their message, the men disappeared. Outlaw had no idea who they were, but he figured they were members of something like Tiny's gang in Connecticut—a generic group of white prisoners, organized loosely around conceptions of white power. He couldn't have been more wrong. Within a day or two, through conversations with black and Hispanic inmates, Outlaw discovered that his visitors were members of the Aryan Brotherhood, the white supremacist gang, considered by the FBI to be the most ferocious gang in the federal prison system. The black inmates told Outlaw that the Aryan Brotherhood essentially ran the prison, and many of the guards were tacitly, or explicitly, complicit in their regime.

Outlaw soon witnessed how quickly the system could victimize new arrivals. He had spent the previous day—the day of endless travel on Con Air—with a black inmate from Iowa, who was soft-spoken, even gentle. They had gotten on the plane together in Oklahoma. When Outlaw and the Iowan had arrived in Kansas, they'd laughed at the craziness of their journey, flying around the country all day, only to land three hundred miles from where they'd started. The first morning at Leavenworth, Outlaw attended an orientation session for new prisoners. On his way to the auditorium, he heard a commotion coming from inside a cell. He looked inside and saw two inmates raping the Iowa inmate. Outlaw later saw him at orientation. He was crying and aloof, standoffish. All of his friendliness of the day before was gone.

In the hours and days that followed the Aryan Brothers' visit, Outlaw desperately considered how he could get his hands on a knife. If the Aryan Brotherhood thought that he was a snitch, they could at any moment return to kill him.

Outlaw had to quickly figure out Leavenworth to get ahold of any advantage he could. He studied the basic geography of the place. The white rotunda was the centerpiece of the prison. Four hallways emanated from the rotunda and led to four cell blocks, each of which had five tiers with shiny metal balconies. A fifth hallway led to the prison commissary, and beyond that lay the ten-acre prison yard, large enough to have subsumed all of Somers. Surrounding everything was the massive wall, which blocked out any view of the prairies. Above the whole monstrosity were six gun towers, where guards with rifles in hand surveyed the hundreds of prisoners milling about below.

Out in the yard on one of his first days, Outlaw noticed a small shack, manned by an inmate who signed out sporting goods—basketballs and footballs, mainly—to prisoners. Outlaw observed that the man in the shack, clearly a Leavenworth veteran, handled himself well. Outlaw approached him, first engaging in small talk about the weather. The conversation was pleasant enough. After a few days of this, Outlaw casually asked the man if he knew how he might be able to get a knife. "If you send five hundred dollars to this address in Detroit, I'll get you a knife," the man said, scribbling the information on a piece of paper.

Outlaw called a former Jungle Boy from the prison pay phone and told him to send a money order to the address in Detroit. A week after that, the man at the recreation shack called Outlaw over. Avoiding the scrutiny of the guards, in a quick and practiced gesture, the man handed over two homemade knives, which Outlaw stuffed in his boxers. They were made of hard, durable plastic, likely from the underside of an industrial chair, and would pass through metal detectors throughout the facility. He felt slightly better prepared as he waited for the Aryan Brothers' inevitable return.

In the interim, he received a letter from Connecticut, from the Interstate Compact office:

Mr. William Outlaw CT #156049 Fed # 45772-006
USP Leavenworth
1300 Metropolitan
Leavenworth KS 66042

Dear Mr. Outlaw:
You were transferred out-of-state for the following reasons:
As part of an ongoing investigation by the Connecticut Department of Correction Security Division and information obtained from outside law enforcement agencies, there was reason to believe you were involved in narcotic trafficking on the streets and within the correctional facility. Per the Commissioner of Correction, you were transferred to the federal system. The Federal Bureau of Prisons will only accept a small box, 15" x

12" x 10" (2 cubic feet) of property. As such, please write the property officer,
at CCI-Somers . . . and indicate what type of property you want sent to
you in this box (i.e., photographs, legal documents, etc.). Please also include
the name and address of an individual to whom you would like your
remaining property shipped.

Outlaw threw the letter away. As far as he was concerned, the Connecticut Department of Correction had tried to kill him in the last few weeks, and he wanted nothing more to do with them.

Three weeks after their visit, the four Aryan Brothers returned to Outlaw's cell, just as he knew they would. They happened to arrive when Outlaw's cell door was open, and he was vulnerable to an ambush. The Brotherhood's leader held a large paper bag in his arms, which he extended to Outlaw. Outlaw wondered if this was a trick, but he felt he had no choice but to peer inside. He saw that it was filled with toiletries— shampoos, toothpaste, soap, all rare and valuable commodities in prison. Outlaw was speechless—the bag seemed to represent a kind of gift.

"Tiny sends his regards," the leader said.

Outlaw, in a flash, realized what must have happened. *The Aryan Brotherhood, or people in their network, must have talked to Tiny Piskorski back in Connecticut.* And Tiny had presumably said commendable things about Outlaw: that he was a straight-up inmate, a heavy hitter, a shot caller. Outlaw could just see it: Tiny, in the Somers visiting room, talking to some kind of intermediaries of the Aryan Brotherhood: "Outlaw's a motherfucker, sure, but he's also a straight-up gangster. He's the real deal." Perhaps Tiny had even mentioned the revolver in the yard that day.

That is surely what transpired. As Outlaw learned more about the operations of the Aryan Brotherhood, he discovered that the gang had a national web of communication among white supremacist inmates, often conducted through their attorneys. Most likely, the Aryan Brothers in Leavenworth had talked to their attorneys—themselves no doubt white supremacist sympathizers—who contacted lawyers who were part of the same network in Connecticut, who interviewed Tiny at Somers about

Outlaw and then sent word back to Kansas. The process had taken about three weeks.

Outlaw accepted the toiletries and said to the Aryan Brothers, "Yeah, I'm one hundred percent official." It was his way of saying that he was legit, not even remotely a snitch. The phrase came to him in the adrenaline rush of the moment, but it would serve as a kind of talismanic slogan that he would use for the rest of his life.

Their message delivered, the Aryan Brothers left. Holding the gift bag, Outlaw felt enormously relieved: he had survived his first major test. But then he heard the sound of their boots coming back down the hall. *Maybe the whole thing about Tiny had just been a setup?* His heart went to his throat.

The lead Aryan spoke again. "One more thing. You might want to find another unit to live in. You should talk to the corrections officer about transferring over to the nigger unit."

THE ARYAN BROTHERS were not incorrect in inferring that the units at Leavenworth, at least unofficially, were segregated by race. The unit Outlaw had been assigned to was dominated by the Aryan Brotherhood, while, he was to learn, the unit on the other side of the tier was controlled by the D.C. Blacks, a major African-American gang. (The D.C. Blacks, composed primarily of African-American prisoners from Washington, D.C., exert a disproportionate influence in the federal prison system. No jails or prisons exist in the District of Columbia, and therefore all sentenced criminals are placed in federal facilities.) But Outlaw had no intention of moving across the tier. He'd seen enough of Leavenworth to know that the Aryan Brotherhood's unit was cleaner and better maintained than the black unit, and that the white inmates received much better treatment from the guards. From that night on, Outlaw actually got on quite famously with the Aryan Brothers. A respect now existed between them, and a kind of détente evolved. Each day Outlaw reflected gratefully on how Tiny had vouched for him. "Tiny Piskorski saved Outlaw's life at Leavenworth," Mike Lajoie said many years later. "There is no doubt in my mind."

Outlaw's paperwork did eventually arrive at Leavenworth. Outlaw believed it to be a strange oversight by Connecticut officials, who, after all, had been plotting his expulsion for months. Perhaps Outlaw was receiving some unofficial payback for all the havoc that he had created back home: arguably, it was just another form of "bus therapy."

Sometime after receiving the unexpected gift bag, and armed with the knowledge that the Aryan Brotherhood, of all groups, had his back, Outlaw stormed into the TV room that was used by the black prisoners. He was furious at them. In his first few weeks at Leavenworth, he had run into some inmates whom he'd met before, even partied with in D.C., Atlanta, Philly, and Baltimore, when the Jungle Boys had roamed up and down the East Coast. But here at Leavenworth, knowing that Outlaw had no pull or political capital out in Kansas, they'd given him the cold shoulder when he'd arrived, pretending they didn't know him.

A football game was on TV, and a group of prisoners, including ones who'd shunned Outlaw, were watching it.

"What the fuck you motherfuckers watching the football game for?" Outlaw yelled. "Don't you know it's the first game of the NBA season, you idiots!" Outlaw didn't care about the football game, or the basketball game, for that matter. The conceit was simply his way of saying: *I've arrived here, I've got some backup, and don't fuck with me anymore.* The bluster worked perfectly. Shocked by his hubris, the black inmates respected him after that.

It must have been extremely tempting for Outlaw to join a black gang, like the D.C. Blacks. He would have been a valuable member given his physical stature and his strategic acumen. Being part of a gang would have afforded him, temporarily anyway, protection and fraternity. Most people in Outlaw's position would have jumped at the chance. But Outlaw never even considered the option. To him it made no sense to join a gang he hadn't started. He would, however, socialize superficially with black prisoners and the D.C. Blacks in the yard and in the mess room. He took to calling his Aryan Brotherhood–populated unit "the Hollywood Hills" because it was at the top of the block and the cleanest, and most gleaming, in all of Leavenworth. At the end of a meal, he'd proclaim to the bewildered black inmates, "I'm going back to the Hills. See you later."

Over time, Outlaw got gradually used to the routines of Leavenworth: five-thirty wake-up call, rec time, breakfast, lunch at noon, dinner, getting locked in the cell for the night at seven-thirty. Still he was inwardly traumatized by what the system had done to him in the last few months, and resolved that the only way to endure was be as stoic and intimidating as possible. He refused all services and work programs, believing that accepting them would be tantamount to giving in to the authorities, and thereby reflect weakness. Instead he played basketball, football, and baseball, and lifted weights indifferently. Massive as he was, he didn't feel the need to bulk up like many inmates did. He also took to selling stamps in prison, a critical part of the underground prison economy. Stamps are considered legal tender, and he used them to buy and sell things—food, radios, sneakers, toiletries—for profit. It kept him busy but inevitably one of his customers wouldn't pay their debt, and Outlaw felt he had no choice but to retaliate. But unlike at Somers, Outlaw didn't have a crew under him at Leavenworth, so he would have to deliver the retribution himself. Usually a few quick stabs in a stairwell or a toilet was all it took. The violence didn't bother Outlaw at all; to him, it was just what one had to do to survive in a brutal environment.

Given his lack of engagement in any of the services offered by the prison, one of Outlaw's few activities during the day was to stand in the entrance hallway in the late afternoon and observe the newly arrived inmates as they entered the facility. For many Leavenworth long-termers, this was an opportunity to assess new blood and figure out who might be easy prey. As the usual group of petrified newcomers filed in one day, Outlaw thought he saw a familiar silhouette. Stocky, short. *No! It couldn't be . . . But it was. Yes! Rodney!* Outlaw ran out into the hall, the guards be damned, and just about leapt into Rodney's arms. He hadn't seen him for years, not since the night of Sterling Williams's murder. The two hugged each other in the prison hallway until a guard told them to knock it off.

Unlike Outlaw, Rodney had been sent directly to Leavenworth from Connecticut. Since the FBI had led the massive bust of the Jungle Boys, he'd been assigned to federal prison. Rodney was placed across the bay from Outlaw, in the (unofficial) black unit, but as much as possible Outlaw and Rodney coordinated their schedules around shared time in the yard and at chow. Even with the Jungle fifteen hundred miles away, it was like nothing had changed between them. With his stout frame and frenetic edge, Rodney handled himself capably in prison. Outlaw was proud of him, as if he were a protégé: they were two Connecticut Yankees who could handle themselves with the big boys.

Sometime into his tenure, Camile flew halfway across the country to visit Outlaw. She was looking forward to seeing him, but when she arrived in the visiting room, she saw a vacant man. "What's the matter, Juneboy?" she said. "What's going on?" He shrugged. He was unfriendly. She couldn't believe his attitude, particularly given the sacrifice she'd made to get there. Outlaw wanted to express his shame and humiliation about his new status to her, but he was unable to. He simply couldn't articulate his vulnerability. Instead his emotions came out as rage. He accused her of being unfaithful. He said he'd heard all kinds of rumors from Rodney and other Jungle Boys with whom he'd been in touch. None of it was true, but Outlaw didn't care. Camile stayed for a week, in a nearby efficiency apartment with another visiting wife to keep costs down. She visited daily, but Outlaw never backed down from accusing her of infidelity. After a while, there was not much for them to say to each other. She could have gotten angry, but given her typical forbearance, she just thought to herself: *He's traumatized, he's in a different world now, let him be.* But on the long flight home to Connecticut, she worried about how they could possibly stay married.

Outlaw called home periodically from pay phones in the hall of the unit. Above the din of other prisoners talking, Outlaw occasionally talked to Pearl. She was still distant, as if she was waiting for something from him, some signal of change that she had yet to see. From former Jungle Boys who had somehow escaped the FBI bust, he heard the news from New Haven. Most of it was bad. Since Outlaw had been in prison, both Blue Eyes and Butchy had died. The rumor was that they'd both contracted AIDs from IV drug use. Outlaw was not surprised. But he received another call from Phaedra's mother that shook him deeply. The moment she identified herself, Outlaw knew something was wrong. "My daughter died," she cried. "She died of AIDS from using drugs. People are dropping like flies around here because of that." Outlaw couldn't believe it. Phaedra dead. She had always been so full of life, spilling over with energy. He cried alone in his cell. He felt ill from the shock and was also full of guilt for his daughter Marquetta, who was now seven years old and motherless. Marquetta was now living with Phaedra's mother. Outlaw tried every week to call Marquetta and comfort her, but there was only so much he could do. They barely knew each other.

UNLIKE IN CONNECTICUT prisons, at Leavenworth televisions within individual cells were prohibited. At first Outlaw was disturbed by this and wondered how he could possibly fill all his time. But within a month or two, bored, he took to visiting the prison library, which was immaculate and freshly lit, its gray metal shelves brimming with well-organized books. The librarians were professional and respectful, even kind. In the perennial battle between prisoners and the administration, they seemed to have sided with the prisoners. If the library didn't have a book, the staff ordered the volume and often it would appear in a day or two. In the library's bathroom, Outlaw discovered a window, albeit an extremely narrow one, that allowed for a view of the surrounding countryside. Outlaw peered for minutes through the window to catch some glimpses of Kansas, to actually have a sense of where he was. One lucky day Outlaw could see deer grazing on a field. It was a thrilling sight that somehow made him feel human again.

With unexpected access to information through the library, Outlaw decided he should learn as much as he could about the federal prison system, with the intent of mastering it just as he had Somers. Who invented it? When did it start? What kind of prisoners was it designed to manage? He ordered all the books he could about the Bureau of Prisons, even a book about Leavenworth, *The Hot House*, a best seller published just that same year. Outlaw read in *The Hot House* that inmates had been forced to build the original prison walls as part of their punishment, and how, in 1931, Leavenworth inmates smuggled revolvers into prison and kidnapped and shot the warden. So—Outlaw smiled—he was not the first to smuggle a gun into prison. He also read everything he could about the Aryan Brotherhood. The gang was formed in the San Quentin prison in California in the 1960s, ostensibly to help white prisoners protect themselves from Latin and black prison gangs, but quickly devolved into a neo-Nazi group. The gang's slogan was "Blood in, Blood out," which meant, theoretically anyway, that the only way to achieve full membership was to commit murder, and, in turn, the sole exit from the gang was death. Being held in a special cell in the basement of Leavenworth at the exact moment that Outlaw was reading *The Hot House* was one of the group's seminal members, Thomas Silverstein. Silverstein had committed two murders of D.C. Blacks within Bureau of Prisons facilities. He was being housed in a special isolation unit made of steel under a "no human contact" rule, the harshest conditions allowed for any prisoner in the United States. The lights were on twenty-four hours a day, and two video cameras tracked Silverstein's every move.

But mainly during his first year at Leavenworth, Outlaw resumed his ongoing project in prison, now in its fourth year: to punish the system that confined him. He might not have been sentenced to eighty-five years anymore, but he still wasn't going to give in. One day on the unit with two white inmates, the conversation turned to prison work programs. "They are a goddamn modern form of slavery. I'm not working for twenty dollars a week making license plates," Outlaw declared. His fellow inmates nodded in agreement.

The next day, December 1, 1993, a corrections officer came to him. "Inmate Outlaw, you are being charged with inciting a riot," he said. "You are being charged with encouraging refusal to work and mandated to disciplinary segregation for thirty days."

"What the fuck did I do?" Outlaw said.

"An inmate said you were causing an insurrection against the administration. This is a very serious charge."

"You're gonna believe a motherfucking snitch over me?"

Apparently the officer did, because Outlaw was cuffed and sent to the solitary unit, Building 63, a three-story brick outbuilding a hundred yards from the rest of the prison. He was tossed into a minuscule room. He looked around at his new quarters: the usual white concrete walls, a bed, a sink, a toilet, a thick metal door, a narrow slot to put the meal tray through, a window to the yard made with frosted glass and surrounded by a cage of narrowly spaced iron bars. Building 63 was one of the oldest buildings at Leavenworth, built in 1898, even before the main prison. It once housed the Birdman of Alcatraz, and in an area behind the building Carl Panzram, a serial killer who killed a guard, was hanged. Just as when he was in solitary at Lewisburg, Outlaw heard the screams of fellow prisoners. Many of these men were mentally ill, howling as if they were being tortured, and responding to internal voices and hallucinations. Outlaw was allowed one hour a day outside of the cell. The guards would jam a window open in the winter to freeze the place, and turn an inmate's allotted shower time from ten minutes into thirty seconds. They would put on the handcuffs too tight, to cut off circulation, and spit on the food tray before passing it through the slot, and call him all kinds of racist epithets. But worse than any of that was the flooding. The toilets would back up and fill the entire floor with water two feet high and swimming with feces and urine. It would often take two days for the maintenance crew to appear. Outlaw would crawl to the highest bunk, and try to wait it out.

Once in solitary, Outlaw accumulated more tickets, which added months to his original thirty-day stay. Below are his charges and his punishment. *DS* stands for "disciplinary segregation":

1/9/94 Being Insolent to Staff member DS 15 days
Outlaw called an officer a racist term.

1/23/94 Being unsanitary or untidy DS 15 days
In Outlaw's view, a nonsense charge.

2/5/94 Assaulting without serious injury DS 30 days
Outlaw threw urine at an officer.

2/7/94 Being Insolent to Staff member DS 15 days
Outlaw called the officer a "racist motherfucker."

2/7/94 Threatening Bodily harm DS 45 days
Outlaw said to staff "next time it will get physical."

2/15/94 Being insolent to staff member DS 21 days
Outlaw mouthed off to staff.

2/18/94 Engaging in Group Demonstration DS 45 days
He once again went on a tirade about prison work programs.

Date unclear Being Insolent to Staff Member DS 21 days
Another threat.

4/16/94 Being unsanitary or untidy DS 15 days
Another ill-defined charge.

4/30/94 Being insolent to staff member DS 21 days
Another insult.

6/14/94 Threatening bodily harm DS 45 days
Another physical threat.

Altogether the charges amounted to 290 days of "disciplinary segregation." Outlaw did what he always did in the midst of a crisis: he slowed things down as much as he could, and coolly considered the options available to him. He was allowed pencil and paper, and he wrote apology letters to Camile. He said he wasn't himself after all he'd been through, and he just hadn't been able to admit to her the shame he felt about not being able to be a husband to her in any way. He thought of all he had lost—he was twenty-six years old now, and had spent six years in prison. He thought of his mother, his children; he thought of the Jungle Boys; he thought of the New Haven Green; he thought of the six-thousand-dollar-a-seat NBA All-Star games; and he thought of the place he would bunk down under the highway as a kid to escape his prowling father. He thought of fishing, watching football, fucking women; he thought of magical Florida highways; he thought of getting drunk and high; he thought of typical New England things, things he didn't even know he cared about, like fall leaves or eating apples. He thought about driving a car on a freeway, or eating a steak and fries and drinking scotch at a restaurant. Or cutting bags of coke with the Jungle Boys. Counting the money. Figuring out the security detail for the day. Getting them to work like a single team, like a coach would. Going to the movies. Holidays: turkey at Thanksgiving, snowball fights, drinking at New Year's. Being the center of the action. Being the man. And all the things he had lost: the money, the cars, the houses, and the satisfaction—the joy even of seeing the fear in people's eyes when they saw him. He never once blamed himself for what he'd done, the violence, but mainly in those dark, unendurable hours Outlaw thought of his five kids, and his mother. He imagined the children grow-ing up, learning to walk and talk and read and run. He had seen none of it. He imagined his mother toiling away sixteen hours a day while he did nothing in a cell.

As the stint in solitary confinement stretched on, Outlaw began to suspect he was suffering from a generalized form of trauma. He could only read so many westerns from the book carts, do so many jumping jacks and sit-ups, and walk the seven feet of his cell back and forth so many times. Social and sensory deprivation left no place for his mind to go but inward. He was like a fish who had been taken out of its natural environs, washed up onshore, flailing away futilely trying to return to the sea. At the end of the day, after the ludicrously early dinner, the dark night of the soul began. Often he just lay in bed with his hands over his ears, trying to pretend that he was anywhere but Kansas. He still only slept a few hours a night. The master manipulator had met his match. *They are calling me on my bullshit. They are outfoxing me, overpowering me, outresourcing me.* It had not been the inmates who had done him in, after all. It was the system, which was like an army that never stopped.

Nights and days became essentially the same. The only real difference was that the frosted glass window was slightly illuminated and warmer to the touch during the day. He began to see that he was just an animal now, no different from a chicken or a calf in some horrendous meat-processing plant. His only purpose in society was to be separated from it, and tre-mendous resources had been spent in order to remove him from normal people: people who worked, people who loved their families, people who paid their taxes and took care of the things around them. He had done none of that, really. Over the course of weeks, his mind seemed to separate from his body, the two wrenched apart like a mangled car. He wondered who he in fact was now, beyond a quivering body somewhere in a basement in the middle of Kansas. No one, except for Pearl and Camile, probably cared for him anymore, and he began to think that he might be better off dead. But the idea of suicide itself never occurred to him: he would not give up like that. Everything became a sleepless canvas of painful, idiotic images, but eventually the funhouse that his mind had become did after a while settle on one horrible enduring image: it was the face of Sterling Williams right before Outlaw killed him, a look of calm almost, as if the man had no idea about what was momentarily going to happen to him. And then he saw the pool of blood that seeped out of Williams right after he fell. He was never going to erase that from his mind, ever.

One afternoon, or at least he thought it was the afternoon, Outlaw heard a familiar cry from outside. He put his ear toward the window.

"Hey, Juneboy! You all right?"

Who could be calling?

"Juneboy! You all right?"

It was like a call from a distant shore, or another continent. Of course! It could only be one person. Rodney! Rodney, out in the yard.

"Rodney! You all right?" Outlaw called back.

"Yo, I'm all right!" came the response.

"I'm all right too!" Outlaw lied.

This became a daily routine, something to look forward to—this call and response with Rodney. It went on for weeks but then one day it just stopped. Outlaw learned later from Rodney that a guard saw their interactions and put an end to it.

One resident of Building 63, named Dutch, had been in solitary longer than anyone else. Given this seniority, Dutch was allowed out onto the galley—the open hallway that connected all the individual cells—to pick up food trays that were rolled in on carts and then distribute them to each inmate. In one of those moments, Dutch whispered to Outlaw that he might be able to get a radio for $250. Outlaw wrote to an old Jungle Boy once again and asked him to send the cash to an address of a friend of Dutch's. A few weeks later Dutch smuggled in a radio on Outlaw's tray. It was a small, cheap transistor radio, but for Outlaw it became a lifeline to the outside world. Keeping the volume low so as not to attract the attention of the guards, he listened to the Kansas City radio stations. He paid attention to politics for the first time, and listened to Kansas City Royals games, and Bible stations from Oklahoma, the weather from Omaha. It didn't matter as long as it brought him to a different place.

In a desperation move he wrote a letter to authorities in Connecticut asking that he be returned back to Somers. On June 20, 1994, he received this curt response:

> Dear Mr. Outlaw,
>
>   Your request to return to Connecticut has been reviewed. The Department of Correction will not return you at this time due to the reason for your initial transfer and your institutional adjustment in the Bureau of Prisons. The Bureau of Prisons has advised you have received 58 incident reports, many resulting in disciplinary segregation. . . . Inmates incarcerated out of state can request to be reviewed on an annual basis. I suggest you make this request in a year. As your institutional adjustment will be a factor, I suggest you try to remain discipline free.

But finally in September 1994, after he had accumulated nine months in Building 63, he was told to meet a senior corrections officer. The man said, "Mr. Outlaw, your time is up. You're out of solitary." But there was a catch. He was going to be sent the next day to Lompoc, California, a "disciplinary transfer" for his lack of progress at Leavenworth. Outlaw knew the name of the facility but nothing else. Still, the idea of California sounded promising. He had never been to the West Coast. If the Feds were going to fly him there on their dime, perhaps that wasn't the worst thing. In the early morning hours of September 20, 1994, Outlaw was escorted out of Building 63 for the last time and shunted into a van. Exiting the Leavenworth compound, the van drove past a huge steel gate. Etched on a sign next to the gate was the following inscription:

LEAVENWORTH PRIDE

Proud of where we have been

Proud of where we are

Proud of where we are going

Proud of a job well done.

———————

COURTESY OF CON Air once again, Outlaw—after a night at El Reno prison in Oklahoma and Phoenix prison in Arizona—arrived at the Santa Barbara airport on September 22, 1994, almost two years to the day after he had been shipped out of Connecticut. On the hour bus ride to Lompoc on Route 101, Outlaw caught glimpses of the Pacific Ocean and its turquoise waters to the left, and the rolling sandy-colored hills of Los Padres National Forest to his right. It was almost hallucinogenically beautiful after nearly a year in solitary.

Arriving at the Lompoc compound, Outlaw saw that the grounds had all the trappings of a federal prison, with the guards wearing exactly the same Bureau of Prisons uniforms, but he could see immediately this was not a historic prison like Leavenworth or Lewisburg. Lompoc was a series of concrete boxes built in 1970, a kind of West Coast Somers, and about the same size: sixteen hundred prisoners. As he went through the intake process—old hat at this point—he sensed, or hoped, that Lompoc would not be as terrifying as those other prisons. Lompoc seemed to be a lesser cog of the Bureau of Prisons machine. Indeed, just five years before, Lompoc had been a minimum-security facility, where Watergate conspirators such as H. R. Haldeman had spent ridiculously soft time. Inmates could wear shorts, guards sported neckties, and the salad bar was known to be excellent.

Still, Outlaw wondered if his life would be threatened within the first hours on the unit, as it had been at Leavenworth. He could tell instantly Lompoc was run by the Hispanics, and in particular the Mexican Mafia gang who walked the halls as if they owned the place. The Mexican Mafia were known for exacting discipline, unrelenting levels of violence, and their influence on the streets of Los Angeles from prison. To survive Lompoc, Outlaw knew he would have to coexist with the Mexican Mafia just as he had done with the Aryan Brotherhood. But he didn't know Spanish nor much about the topography of West Coast gangs generally. And he would have no Tiny to vouch for him. To complicate matters further, the black gang landscape too had been transformed in the time he had been away. Two rival gangs—the Bloods and the Crips—had risen from the streets of Los Angeles. The Bloods and the Crips had flooded Lompoc, and there were constant fights in the TV rooms between them.

On Outlaw's second day at Lompoc, a Crip rushed up the stairs outside Outlaw's cell. "Heads up!" he shouted. Outlaw heard yelling and commotion coming up from the stairwell. Suddenly three Hispanic inmates stood outside his cell, their chests heaving from the exertion of running up the stairs. *Here we go again,* Outlaw thought, preparing for another threat to his life.

"You are Mr. William Outlaw?" one of the men asked. He had a strange, almost reverent tone in his voice.

Outlaw nodded, guardedly.

"Oh, that's great." The Hispanic prisoner smiled. "We can't believe it's actually you." He held a large plastic bag. Outlaw thought this might be a ruse, and they would jump him—kill him—in seconds. The inmate extended the bag to him. Inside was a half pound of marijuana and fresh apples and oranges.

"Our friend Termite sent us," another one of the group continued pleasantly. "He wanted to thank you for everything you did for him. He really appreciated it. These are from Termite." Outlaw broke up into laughter. Termite! Termite had been an inmate on Outlaw's block at Leavenworth before he was sent to solitary. Termite, a diminutive but muscular guy, was upbeat and chatty, and exuded an impressive energy. Outlaw had understood from the jailhouse chatter that Termite was a major player in the Mexican Mafia. Termite called Outlaw *gordo,* meaning "fat guy" in Spanish. One day Termite had approached Outlaw and sheepishly asked if he might possibly do him a favor. Termite had a box with him, and asked if Outlaw could send it out for him to his family back in California? Termite said it contained nothing illicit, just a box of personal photos and letters. For the postage, Termite handed Outlaw forty dollars, which, he explained, was all the money he had.

"Why can't you send the shit out yourself?" Outlaw asked.

"They don't like me down there," Termite said about the post office workers, who were part of the prison staff. It sounded like an excuse to Outlaw. Still, Outlaw liked Termite and mailed the box to an address in Los Angeles. The bill came to $52. A twelve-dollar difference: a not insignificant sum in prison. Outlaw paid it with profits from his stamp business. Back on the unit, Outlaw explained what had happened. Termite couldn't thank him enough. "You cool, Outlaw. You the best." Shortly after that Outlaw was sent to Building 63.

Now, outside Outlaw's cell at Lompoc, the Mexican Mafia members explained that just after Outlaw had gone to solitary, Termite had instigated a riot against the Texan Whites, a rival white gang in Leavenworth. A number of the Texan Whites were stabbed. Termite had been planning the insurrection all along, and wanted his personal photos and letters sent out before the violence brought him under scrutiny. Indeed, after the riot he was shipped out to Marion in Illinois. The guys from the Mexican Mafia explained that Termite was their superior in the gang, and he personally had sent word to them to take care of "Señor" William Outlaw. "If you have any problems here at Lompoc," they said, "just let us know. We take care of it for you, no problem." With the backing of the Mexican Mafia, who made good on their promise to look out for him, even allowing him to use their TV room, Lompoc became a comparatively good time, at least relative to Lewisburg and Leavenworth. Outlaw appreciated the radiant blue skies and sun in the yard all year long, and the special deliveries of pot and vegetables from his Mexican Mafia friends. Outlaw even ended his moratorium on prison programs and engaged in welding and construction workshops. But at the same time, he continued to deal stamps and receive incident reports.

In the yard, Outlaw noticed one inmate, five feet six inches tall and

150 pounds, who spent time growing white roses and tomato plants in a garden. The man was in his sixties and Outlaw saw that he had a freedom of movement within the prison that no other inmate was afforded, and that many of the officers were obsequious to him.

"You from Philly?" the man said, spotting Outlaw's erroneously assigned inmate number. The man spoke softly, but Outlaw thought he could detect a raspy, Brooklynesque edge in his accent.

"No," Outlaw said, "New Haven."

The man brightened, and introduced himself as Carmine Persico. "I'm from New York. What the hell are two East Coasters doing all the way out here in California?" He laughed. Outlaw and Persico talked about the weather, growing tomatoes, nothing of great import. The next afternoon in the yard they chatted some more, and the next day, some more again. One day, Persico mentioned something about being involved in the Mafia.

In the prison library, which was run with the same professional precision as at Leavenworth, Outlaw ordered books about the Mafia. He was astonished—but not entirely astonished, as nothing in the federal system surprised him anymore—to read that Persico had been the boss of New York's Colombo crime family since 1973, even though he had been imprisoned for all but four years during that period. Persico was running the family from his prison cell, three thousand miles away from New York City.

As a teenager in Brooklyn, Persico gained the attention of Mafiosi when he killed a rival with his bare hands. Initially hired by the Profaci family to assist with their numbers operations, Persico emerged as a valuable enforcer and rose within the mob. He became known as "the Snake" for his tendency to switch sides based on political advantage. Yet he was also oddly loyal, a dutiful husband and father. After narrowly escaping death during a drive-by shooting, he refused to identify the man who shot him because to do so would violate the rules of the street. Persico proved so ruthless and brilliant that he took over the Colombo family organization, even though he was originally an outsider. After being indicted on racketeering charges in 1984, Persico went on the run, landing on the FBI's Ten Most Wanted list. Four months later, he was caught and eventually tried. While he had retained lawyers in the past, Persico believed he had sufficient legal experience to defend himself this time. The scheme backfired and in 1987 Persico was sentenced to 139 years. At the sentencing, the judge told him, "Mr. Persico, you're a tragedy. . . . You are one of the most intelligent people I have ever seen in my life." After the trial, Persico put out contracts for the murder of the lead prosecutor, Rudolph Giuliani.

Years later, a former inmate who knew Persico well confirmed that Persico ran the Colombo family while incarcerated by sending messages through visiting lawyers and relatives, bribing prison guards, and frequently approving mob hits. Persico admitted to his confidant that he was responsible for the killing of twenty-five men, half of whom he had murdered himself. But to Outlaw, Persico was a gentle older man, surrounded by his roses and tomatoes. He took to making homemade pasta sauce for Outlaw, and they became friends. No doubt Persico recognized in Outlaw an East Coast criminal talent like his own.

Just as things were starting to finally feel bearable, the Feds abruptly told Outlaw that he was being sent back to Lewisburg. He had been in California only eleven months. "Lewisburg? Do you have to send me back to Lewisburg?" Outlaw cried to the unit manager. It would be the only time during Outlaw's entire prison odyssey that he found himself actually pleading. As he was flown across the country once again, with overnight stops in Phoenix and Oklahoma, dark thoughts entered his mind. He felt that returning to the ghoulish dungeon of Lewisburg was going to overwhelm him and take over his soul. He wouldn't be surprised if he died there.

Part Two

CITIZEN

The darker the night, the brighter the stars.

—Fyodor Dostoevsky, *Crime and Punishment*

CHAPTER 6

How to Escape Your Prison

"Doc" Whitmire, Lewisburg.
Courtesy of Richard Whitmire.

UPON READMISSION TO LEWISBURG ON AUGUST 14, 1995, OUTLAW, NOW
twenty-seven years old, was assigned to A Block, which turned out to
be the most notorious unit in the prison. A Block was like a strange
dark castle: everything was made of stone and red brick, the hallways
were narrow and mainly windowless, and the stairwells jutted out at odd
angles. The block was comprised solely of single cells, an extremely unusual arrangement in the Bureau of Prisons. All inmates on the unit had
serious criminal pasts, and most had life sentences. But strangely, Outlaw, initially at least, found A Block to be relatively peaceful. He found
his block mates to be more secure in their personas than typical inmates,

perhaps having less of a need to prove themselves. The inmates were much older than was typical too; in fact, Outlaw was the youngest among sixty men. The other oddity was that A Block was an open unit. Inmates could freely congregate in the common areas and keep their cells unlocked to receive visitors. Guards were supposed to make their rounds every fifteen minutes, but often they didn't appear for an hour at a time. It was almost like living in a dorm—which, of course, Outlaw had never done.

In order to maintain his street cred, Outlaw resumed selling stamps. One day after an inmate hadn't paid a five-hundred-dollar stamp debt, Outlaw stood at the bottom of a stairwell holding a shank, waiting for him. He heard the steps of the man as he descended the stairs. The inmate, nicknamed Shorty, was five foot five. The moment Shorty saw Outlaw he froze, and in the next instant Outlaw grabbed him in a headlock and plunged the shank into Shorty's rear. Outlaw pushed the knife in and out: once, twice, three times. Shorty fell to the floor, and Outlaw ran up the stairs to his cell, where he washed the knife off and slipped it under his mattress. He was unconcerned about any repercussions. There was no way Shorty was going to report him after the maiming that he had just suffered.

Later, in the yard, an older inmate lashed out at Outlaw: "Why'd you have to stick it to Shorty?"

Outlaw yelled back, "I don't care if the dude was a giant, a midget, an Indian, a black dude, a white dude, he owed me money! I wish he'd been the biggest dude in the prison. I have to show everybody what happens when you don't pay!" Outlaw felt like he needed to keep up his smoldering, seething image at all costs.

But then, oddly, almost inexplicably, things began to change.

EIGHT MONTHS INTO his second admission to Lewisburg, Outlaw received a phone call from his daughter Marquetta, who was now ten years old. Outlaw had not seen her for four years, since he'd left Somers. She attended Catholic school in New Haven, where she was a good student, but wanted to transfer to the public school because that's where most of her friends were.

Outlaw said to Marquetta, in his definitive way, "You should stay at St. Mary's."

"Why, Daddy?"

"Simple. It's a better school."

"But, Daddy, did you finish school?"

"No, I didn't."

"Oh," Marquetta said, sounding surprised.

In the awkward silence that followed, Outlaw felt as if he had been shanked. Actually, he would have preferred to have been shanked. He fumbled with his words as he continued to speak to Marquetta, and tried to get off the phone as soon as he could. He struggled to walk back to his cell, crawled on top of the mattress, and looked up at the white, blank ceiling, his mouth tight and a tear forming in his eye. It was as if a dam had broken inside him. He was a fuckup, not even capable of giving a little girl advice, a girl whom he had fathered but had never supported with even a single dollar. Of all the things he had been through in the last decade—the Interstate Compact, the Aryan Brotherhood, nearly a year in solitary—this was the worst feeling he'd had to bear. He wanted to howl.

OUTLAW DIDN'T SLEEP at all that night. At the first opportunity the next morning, he signed up for a high school general equivalency degree class, which met three nights a week. Later that day he stole the G.E.D. instruction manual from the library and studied rabidly every night inside his cell until he got the degree, with ease, three months later. When he got the certificate, on August 12, 1996, he made copies of it and sent the original to Marquetta. Only after she received it did he have the self-respect to talk to her again.

On January 2, 1997, still flush with pride in his achievement, he sent an ebullient letter to the Interstate Compact Office in Connecticut.

Dear Ms. ——,

This correspondence is provided for your evaluation of my progress since Connecticut relinquished custodianship to federal authority.

While my coming to federal custody was begrudging, the stay has been anything but.

Although the proverbial High School diploma is considered a miniscule achievement for some, my self-esteem really soared the day my name was finally etched upon this grammatical cloth of eternal announcement.

If I should sound like a role model it is because I am hoping to really become one.

Very truly your's,
William Outlaw

The letter was not actually written by Outlaw, but rather by a friend, a jailhouse lawyer type who offered to compose it and evidently specialized in florid language, but the sentiments were real. Being exposed to that first taste of education was transformative to Outlaw. He enjoyed all the studying—writing, solving math problems, and reading history—the first schoolwork he'd done for thirteen years. For the first time since the early, heady days of the Jungle Boys, he had proved that he could do something creative, that he could produce something, and not just do damage. It was like a window opened inside him, and he vowed to himself that he would pursue an education whenever he could in the future.

At around this time, Outlaw was befriended by an inmate on A Block named Frank James. James was well into his fifties, and one of the few gangsters in America who, in his prime, was more merciless than Outlaw. A former junkie and stickup artist, James was one of the original members of The Council, New York City's and America's biggest heroin gang of the 1970s, formed when the seven major heroin dealers in Harlem banded together to become one supergang. The initial semi-idealistic vision of The Council, reflected in its name, was that its seven lead members would act as a kind of crime consortium, voting democratically on policy decisions. In the few photographs that exist of The Council, Frank James is strikingly handsome and athletic-looking, appearing more like a safety for the New York Giants or a point guard for the Knicks than a killer. The Council's leader was a flashy sociopath named Nicky Barnes, who, knowing the capabilities of his colleagues well, named James as the gang's enforcer.

In 1977, the *New York Times Magazine* ran a cover story on Barnes: "*Mr. Untouchable: The Police Say He Is Harlem's Biggest Drug Dealer, but Can They Prove It?*" Barnes was billed as the top drug dealer in America and proud of his ability to elude the law. With his debonair clothes and expensive cars and countless women, Barnes became the archetype of the "original gangster." He had five homes and six bodyguards. President Jimmy Carter was said to be so outraged by the *Times* article that he personally ordered the FBI to engage in an all-out effort to convict Barnes. But when New York City and federal law enforcement officials embedded themselves in the community to track him, they discovered that Barnes was deeply popular. He had styled himself as Harlem's Robin Hood, giving cash out in the community and serving as the deacon of his church, delivering turkeys on Thanksgiving.

But the true credo of The Council was found in its motto: "Anyone who is in power who is not willing to terminate, will be terminated." Of all the members of The Council, Frank James was the most willing to terminate others. Even when he didn't pull the trigger, James was central to the gang's enforcement activities. When Barnes was eventually caught, he blithely became a government informant, and he divulged that James had murdered his brother-in-law with an ice pick and killed a suspected snitch by shooting him in the face. In 1984, Frank James was sentenced, at age forty-four, to life without parole, plus an additional forty years.

Outlaw knew all about The Council from gang culture and rap songs. Given the world that he had grown up in, encountering a figure like Frank James was like meeting a superstar like James Brown or Stevie Wonder. He couldn't believe he was on the same unit with him. But the version of Frank James—now converted to Islam—that Outlaw encountered at Lewisburg was quite different from the enforcer of The Council. James was gentle and calm, always sweeping the floors and obsessively cleaning up the unit and stressing the importance of education and faith. When Outlaw arrived on A Block, James showed him around and introduced him to guards and other inmates. He advised Outlaw who was a "stand-up guy" and whom to avoid. Outlaw admired James's sharp, assured manner and they soon became inseparable. They saw each other at five-thirty when the doors of their cells were popped open, ate breakfast, went to the yard, and watched CNN together. In the evenings they would play cards or talk for hours. James was becoming the father figure that Outlaw had never had.

One day, James called Outlaw over to him. "William, I'll never get out of here," he said. "I'm here for life. I'm done. But you're different. You will get out someday. And you're a talented guy, you have leadership abilities. You have to do something with your life. If you don't do it for yourself, then at least do it for me."

No one had ever said anything like that to Outlaw before. When he was younger, he had been too assertive, too arrogant, and at times too savage to ever receive advice. Outlaw was used to giving directions, not getting them. In fact, not since he was about fourteen years old—not since Coach Saulsbury and the Wilbur Cross basketball team—had anyone told him what to do in any kind of positive or productive way. He had only been told what *not to do*, or what he'd done wrong.

You have to do something with your life.

From the moment of that conversation, something inexorably shifted inside Outlaw. That someone of the caliber of Frank James was even talking to him, and actually believed in him, was transformational for Outlaw. In the aftermath of the conversation, Outlaw set goals for himself, goals that were no longer strictly of a criminal nature, or purely manipulative, or about how he might get over on people. For the first time in his life an authoritative male figure had shown absolute faith in him and given him guidance. He was anxious to follow it.

Outlaw suddenly undertook a flurry of positive activity. He completed a parenting program, in which his attendance and participation were commended as being "excellent." A month later he entered a stress and anger management program. In January 1997, he was assigned to the prison hospital to work as an orderly. His performance evaluations reflected a "good to outstanding performance. . . . Both of his supervisors have submitted memorandums which indicate inmate Outlaw is an asset to their department." Outlaw went on to take a continuing education class in economics in August, and after that, basketball and softball officiating classes. A report by staff on Outlaw's conduct from January through June 1997—a year and a half after his readmission to Lewisburg—went as follows: "Inmate Outlaw exhibits a mature and positive attitude, he gets along with both inmates and staff. Over the years, he has worked to change his attitude for the better. . . . Currently he works 154 hours per month as a hospital orderly. [He works] 5 days a week, 6 to 7 hours a day, Monday to Friday. . . . Inmate Outlaw is continuing to successfully program at this facility. His Unit Team is recommending inmate Outlaw for a custody reduction. . . . We anticipate recommending him for a Lesser Security Transfer, to a Federal Correctional Institution closer to his release residence."

The two conversations, with Frank James and with Marquetta, occurring within a few months of each other, acted as a one-two punch to Outlaw's psyche. They forced him to feel shame and loss about the past, but at the same time gave him a thirst for new opportunities. He didn't know exactly where things would lead, but he was committed to starting anew. He was still going to be a "hundred percent official," but in a different way. He wasn't going to manipulate and punish people anymore. Shortly thereafter he stopped selling stamps and went so far as to apologize to Shorty, who was predictably astonished.

It can also not be ignored that around this time he was nearing thirty years old. It is possible that Outlaw was simply "sick and tired of being sick and tired," as former addicts often put it. Being a menace all the time required constant vigilance and energy, and perhaps Outlaw simply couldn't muster that role anymore. With his chronic sleeplessness, perhaps he was simply exhausted after fifteen years of terrorizing people. As the author David Von Drehle has written, "Violence is typically a young man's vice; it has been said that the most effective crime-fighting tool is a 30th birthday."

In any case, one year later, on the night of August 28, 1998, two events occurred on A Block that changed Outlaw's relationship to violence forever.

Outlaw was in the TV room with ten inmates that night to watch a playoff game between the Houston Comets and the Charlotte Sting in the WNBA, the newly formed professional women's basketball league. Early in the contest a guard slammed the door of the TV room shut and bolted it from the outside, locking the prisoners in. Outlaw knew some kind of incident had occurred but went uneasily back to watching the game. Two hours went by, and Houston won the game handily. Sometime later another guard opened the door and tossed a garbage bag into the room, filled with rolls of toilet paper. Outlaw tried to ask what was going on, but the guard locked the door shut as quickly as he had opened it. The group stayed in the TV room until 4 A.M., when two more guards appeared and instructed the inmates to return to their cells. From there, they were called individually to the captain's office and interviewed. *Did they know anything about the murders that had just transpired?* Neither Outlaw nor anybody else in the TV room was aware of what had occurred, but they would find out soon enough.

Sometime early in the basketball game, on the floor below the TV room, three members of the Aryan Brotherhood—Outlaw's old nemesis from Leavenworth—had ambushed the unit from outside and murdered two African-American prisoners, Frank Joyner and Abdul Salaam, and stabbed a third man. Joyner was playing Monopoly when he was killed. Salaam was lying on his bunk when he was stabbed thirty-four times with a twelve-inch knife made of scraps from the metal workshop.

Outlaw had been friendly with both Joyner and Salaam, and had in fact been playing Monopoly with Joyner earlier in the day. To Outlaw, Joyner was an upbeat guy known around the block as an excellent and fair (unusual for prison) referee in the unit's basketball games. Salaam was an exceptional athlete; he would run miles around the prison yard. He had just come in from a run before he was murdered. As Outlaw heard about the attacks, and the coordinated way in which they transpired, he was deeply shaken. He couldn't believe it—he had just been joking around with Joyner and Salaam earlier that day. But at the same time, Outlaw, the old shot caller, couldn't help but be impressed by the Aryan Brotherhood's maneuver, and the level of precision and organization that it required.

Joyner and Salaam were targeted because they were members of the

D.C. Blacks. The beef between the Aryan Brotherhood and the D.C. Blacks went back decades, as all things did with the Brotherhood, to the gang's patriarch Thomas Silverstein, who at the time of the Lewisburg attacks was still in solitary at Leavenworth. Silverstein himself had killed two D.C. Blacks while he was incarcerated. After one of the murders, Silverstein had dragged the corpse up and down the tier for all the inmates to see, some cheering at the sight. In retaliation, D.C. Blacks killed a white inmate at Lewisburg in 1996, and in 1997 a dozen members of the D.C. Blacks attacked six white inmates at a federal prison in Illinois. In response, the leaders of the Aryan Brotherhood decided by the summer of 1997 to "kill whatever D.C. Blacks and associates we could." One of the gang's commissioners, T. D. Bingham, ensconced in a Colorado prison, used a method of secret communication, employed by both George Washington's Continental Army and Pliny the Elder, to get the message to Allen Benton, the leader of the Aryan Brotherhood at Lewisburg. Using an ink of citrus or urine, messages could be written invisibly on paper and later read by exposing the ink to heat. The letter to Benton arrived at Lewisburg in the 4 P.M. mail the day before the killings. To a casual observer, the note was written in standard ink, but Benton, with the help of an associate, lit a match and exposed the paper to the warmth of the flame. The concealed message came into view: "War with DC Blacks, T.D." Bingham's message was, in other words: *Kill every black you can find.* Benton flushed the letter down the toilet and carried out his instructions. He later said in court that he picked Salaam as a victim because they were friends. He wanted Salaam to die quickly and with minimal pain. "I killed him because I liked him," Benton told the jurors. "I made sure he would not have to suffer too much. Something sloppy—it goes on for a while and I imagine it's painful. I hit the right places, vital places so he dies quick."

After the killings, A Block was placed on an indefinite lockdown. All prisoners were confined full-time to their cells, only allowed to go to rec in the yard or the shower room every other day. There was no longer any congregating in common areas. The dorm-like, even at times convivial atmosphere of the unit was long gone. The lockdown went on for weeks and then months. Outlaw did occasional calisthenics and read pulp novels, but that did little to stave off the boredom and the isolation. He desperately missed talking to Frank James and making phone calls home. He missed the classes he had been taking and he missed working as an orderly. Outlaw began to wonder whether he could withstand what was beginning to feel like another extended round of solitary confinement and its attendant desperation. *Maybe,* he thought, *this is what will finally kill me.*

After five months, the lockdown ended. Inmates could once again go out to the yard for a few hours a day, they could eat together, and more important, they could talk and hang out in the common areas. Outlaw and Frank James bounded up to each other and hugged like long-lost family. Emerging from the lockdown, the unit became closer, and bonded around having collectively survived the loss of two of their members.

Outlaw had been around shootings and stabbings his whole life, but the murders on A Block hit him in a way that nothing else had before. Perhaps because he had nothing at all to do with the beef—he was completely unengaged in and unmoved by the decades-old struggle between the two gangs—he was able to see violence afresh, almost as if for the first time. It may have been odd for someone who'd grown up around murder, but Outlaw simply couldn't get over the shock of it. He had just said hello to Salaam on the afternoon of the day that he died, as Salaam came in from the yard, sweating from his run. Outlaw could see both of the victims smiling, laughing; he could see Joyner refereeing a baseball or a basketball game, Salaam running powerfully around the block. It made Outlaw think about the murder he'd committed, and his memory of Sterling Williams. He thought, really for the first time, about what Sterling Williams might have been like as a person—until then, he'd just been a caricature, an enemy, a threat to be eliminated. It was a painful thing to consider. He wondered where Sterling Williams had grown up in Jamaica, what he had been like as a kid. What kind of ice cream he liked.

In the wake of the killings, the administration required inmates who had committed particularly violent crimes to see a psychologist or a social worker. Outlaw was randomly assigned to Richard Whitmire, a seasoned forty-five-year-old social worker who had worked in the prison for seven years. Whitmire was also a commissioned officer in the United States Public Health Service, which provides health care in federal departments. As such, he wore a military-style uniform that resembled navy whites, with navy blue "hardboards" on the shoulders emblazoned with three gold stripes. His rank was that of commander, the equivalent of a lieutenant colonel in the army. In his full uniform, Whitmire cut an impressive, self-assured, almost dashing figure. The inmates called him "Captain" or "Doc," while the nurses called him "Magnum," because of his resemblance to Tom Selleck, star of the popular television show *Magnum, P.I.*

Outlaw was given a special pass to meet with Whitmire in the mental health and medical unit, which adjoined A Block. Whitmire's office, in a former cell, was cramped and dark. Even though the administration had done their best to pretty up the place and give the room a professional veneer, there was no disguising that inmates had once lived there.

At their first meeting, Whitmire affably told Outlaw that they would meet weekly for a period of a few months. Outlaw sneered at Whitmire. He didn't like anything about this forced arrangement. He had not asked for help and had no interest in receiving any. The idea of sharing his feelings with anybody, particularly a man, was absurd to him. Whitmire shrugged and said, "This time is for you, not me. I get paid by Uncle Sam whether you talk or not. Plus I get to go home tonight after I leave here." Outlaw leered at him, picked up the single copy of *Sports Illustrated* in the office, and read about football. Whitmire, nonplussed, used the time to catch up on paperwork. The stalemate went on for three sessions, and Whitmire got a lot of busywork done. But finally, bored with reading the same magazine week after week, Outlaw asked Whitmire a question for the first time. "What do you think of how the Philadelphia Eagles are doing?" "I think they need help on the O line," Whitmire responded instantly. "Me too," Outlaw said, almost in spite of himself, shocked that a social worker could answer such a question.

At the next session, Outlaw asked Whitmire a few questions about the social worker's background. He fully expected Whitmire to parry the questions. To Outlaw, that's what inauthentic people like Whitmire always did. But Whitmire freely shared that he had grown up in blue-collar rural Pennsylvania, as he put it, "in a bar." He explained that his parents owned a saloon on the ground floor of a building and his family lived in the apartment above. "I grew up in the drug business at an earlier age than you did," Whitmire said. Outlaw then asked Whitmire where he went to school. This had become a particular area of interest for Outlaw since he'd gotten his G.E.D. Whitmire didn't mention the college and graduate schools that he'd attended, but he did tell Outlaw that he'd been kicked out of parochial school for defying Sister Rita. Then, seemingly out of nowhere, Whitmire told Outlaw to ask him what his favorite song was. Outlaw told him he didn't give a damn what his favorite song was.

"Come on, ask me," Whitmire insisted, smiling.

"All right, what's your favorite song?" Outlaw said.

"It's 'I Fought the Law (and the Law Won),'" the social worker deadpanned.

Outlaw was starting to realize that Whitmire was a little different. For one thing, he was hilarious. Whitmire told Outlaw he was a specialist in "druggery, thuggery, and skulduggery." He said that he was "just a 'whitey' working in the plantation system, merely helping his clients do a 'checkup from the neck up.'" Whitmire talked to Outlaw as if he were an actual person, a human being, not a "nigger" or an "inmate" or a "motherfucker," as he was used to being called. For almost a decade now, and perhaps for good reason, given his behavior, Outlaw had only been screamed at, ordered around, cuffed, and told what was wrong with him and what he was incapable of. No one beyond other inmates had spoken to him in a humane way for as long as he could remember. The system simply didn't allow for that. Posturing, aggression, and fury were what drove the prison system: if you let your guard down for a second, you could get wheeled out on a gurney. But here in Whitmire's former prison cell office, things felt different. And Whitmire wasn't just a shrink; Outlaw saw how he often talked about practical things: making healthy choices, managing money, the importance of getting an education. Whitmire looked like a million bucks in his commander uniform, worked out, practiced yoga (and even admitted it to the inmates), and he had the nurses on the medical unit doting on him. Not since Mike Lajoie back in Connecticut had Outlaw felt this kind of respect for a prison official.

TENTATIVELY AT FIRST, Outlaw started talking about his life and his childhood. Outlaw talked about Pearl. How he loved her and how he worried he had broken her heart; about how she was still working two factory jobs back in New Haven and hadn't been able to see her son for nearly ten years. Outlaw talked about his five children, who had been babies and toddlers when he first went in, the youngest of whom were now approaching ten years old. He felt terrible guilt about that. As he had gotten older, he had been thinking more and more about his children. He had always stayed at least nominally in touch with them, but now, he told Whitmire, he was going to write to them all the time. He hadn't been there for them, and he felt a powerful sense of urgency and regret. Other people had raised them—their various mothers, as well as Pearl herself. These were things he was not ever going to get back. He desperately wanted to show his kids that being a gangster and a prisoner wasn't the entirety of their father's life story. Whitmire quietly listened, and constantly encouraged Outlaw that it was not too late to make a change.

Eventually in that dingy office, Outlaw talked about his father. He talked about waiting on the stoop for his father with his homemade fishing pole for hours, only to be stood up. He talked about beatings with the belt. The feeling that violence was normal. Outlaw talked about his own searing, nonstop fury: his anger at his father, his grandfather, even at Pearl. How when he was younger he didn't value his life, and therefore he couldn't value anybody else's. Whitmire said, "Until you express that anger, there's nothing I can do for you." Outlaw realized he had been in a rage all these years, immersed in a bitter, poisonous, limitless, mournful emotion that made him want to eviscerate just about everything, from Q View to Building 63 at Leavenworth.

Over time, in his direct and practical manner, Whitmire asked Outlaw to consider that he might be able to change what was right in front of him, rather than what had occurred in the past. "You don't necessarily have to be a product of your environment," Whitmire said. "You could create a new narrative." But how could Outlaw do that? He had already spent ten years wasting away in prison. "You're still young," Whitmire said. "When I was your age I was still . . . well, I don't want to tell you . . ." Whitmire laughed. The entirety of this message was difficult for Outlaw to process, but once he took hold of it, in his typical all-or-nothing style, it was remarkable how quickly his bitterness faded. On an almost daily basis, Outlaw felt a kind of slow melting away of anger. It was a relief not to have to fight everything. He was not especially spiritual, but it did lead him to think about God at times. He wondered if God would, or could, forgive him for his past.

Whitmire was assigned to the care of hundreds of inmates, and one day he respectfully told Outlaw that he could not continue their individual sessions indefinitely. However, he could refer him to his weekly group therapy sessions, held for fifteen inmates and lasting thirty weeks. Whitmire said the name of the group was "How to Escape Your Prison." Outlaw was confused for a moment, thinking the group was actually about how to physically escape from prison, but then he realized that the operative word in the title was "your." Outlaw enthusiastically agreed.

At the first session Whitmire told the group they were going to follow a curriculum in a workbook, which cost $25. He insisted that all participants find a way to pay for the book, so that they would have skin in the game. (Whitmire collected the money and sent the funds back to the government, in a check made out to the U.S. Treasury. The government, unaccustomed to such refunds, often had difficulties processing the payments.) Whitmire also explained that the group was a twelve-step program, like AA, in that participants had to complete one step before moving on to the next. There would be homework exercises outside class. Ultimately, the social worker explained, there were only two requirements of the group but they were both nonnegotiable: honesty and respect.

Outlaw was among the first to buy the workbook and he pored over it in his cell. He read that the purpose of the curriculum was for members "to appraise their current relationships to confront their current and former attitudes and behaviors. . . . Unhappiness was and is part of your life because you chose it." The first of the twelve steps, Outlaw read, was "To begin the escape from the difficulties and problems of your life, and to take control of your life, you must admit that you are the source of the problems in your life. . . . If you can't do Step 1, it is simply because you are not ready to change. It means you are dishonest and not to be trusted. And it also means that you want to stay that way. For the only escape from disloyalty is to quit pretending and to become 'for real.'" Some of the inmates in the group were turned off by how direct and confrontational the language in the book was, but Outlaw liked it. It was up-front and real, just like Outlaw himself.

Outlaw quickly became a leader in the group, completing his homework with maximal effort, and pushing others to do so. A couple of the members of the group were late in paying their $25. Outlaw good-naturedly cajoled them, saying they had to be invested in the process. "Come on, man. Let's go. Let's do this shit." At one of the first sessions, Whitmire asked Outlaw to stand up in front of his peers and give them a completely honest assessment of his problems. He said he wasn't happy with himself, and that he had wasted a lot of time. He couldn't stand it anymore, and he zealously wanted to change. For homework before the next session, Whitmire instructed the group to do an exercise called "The Man in the Mirror." They were told to stare into a mirror for two minutes straight without turning their head away, looking themselves directly in the eye. Outlaw did the exercise in the shower room. He didn't like the man he saw, but he kept his eyes fixed on himself until the very end. He vowed that next time he did the exercise he'd see a different person. A later step in the curriculum asked the group "to begin to repair the injury you have caused to yourself and others." Outlaw was asked to list people in his life he had damaged, and what he could do about cleaning up the mess. Of course he wrote about Pearl. He had tortured his poor, long-suffering mother, as well as countless others, including women he'd taken advantage of and addicts he'd afflicted, leading to their early deaths. In a later exercise, Outlaw was asked to help other people, "especially those who will give you nothing in return," and to document ten hours in which he had done so. Outlaw put in extra unpaid hours in the hospital, wheeling prone prisoners to their appointments and sitting at their bedsides, keeping them company. After these early stages were completed, Whitmire asked the group to move on to

Dear ——

I pray that you are doing well. I am doing fine.

I have made tremendous strides over the past 8 years. I have nothing but the utmost gratitude and respect for the person or persons who sent me into the Federal Prison System. Being here has made me a better man, a better father and a better son. A few years back, I came upon the fork in the road. I've chose the road that requires me to have a good positive attitude, a high moral belief system and respect for God and the Laws that established to govern our society.

Through programs and certified courses, I've sought to address all of my character defects and flaws. . . . I truly look forward to returning to Connecticut in the near future. My successful return will shine a bright and positive light on you and the positive impact that the Interstate Compact Program will have on many of the Connecticut inmates housed out of state. Being one of the early high-profile inmates to return after a substantial amount of time away can only serve to quell many of the doubters of your program.

Thank you and God Bless.

William Outlaw

This letter was followed shortly thereafter by another, also to the Interstate Compact office:

I do not believe in any way that my past criminal behavior was appropriate. I have long accepted full responsibility for my actions, and realize the seriousness, and the harm I've done to not only society as a whole, but my family, including my children. Much pain and sadness has been caused by my actions and regretfully, I cannot turn back the clock and undo what I've caused. . . . Not a day goes by where I don't think of the victims in this case. . . . Each day, I make it a point to remember them all because I don't want to forget. It is the debt I owe them all. . . . It is not just about punishing me for what I've done, but it's about atoning, it's about repairing the damage that's been done to the fabric of life. [My children] live with the stigma of their father's actions, an unfair burden to them. . . . I do not want my children to be ashamed because of what I've done. . . . My mother has had to suffer not only from my behavior, but from my absence. . . . I know she has felt the guilt for my actions, and has asked herself many times 'where did I go wrong?'; yet despite the many conversations we've had about my actions being my fault, she still questions herself. I feel terrible about this and wish that I had listened to her way back when. . . . I want to use my knowledge, my experience, to work with young adults at risk in my community. . . . I want society to know that I have come back from the dead, spiritually and mentally. There has to be a positive ending to the story. . . . I know in my heart that if given the chance, I will be a beacon of light to those in need.

The progress reflected in this letter, and Outlaw's now two years of exemplary behavior, was recognized by the Bureau of Prisons. In July 2000, Outlaw was transferred to a lower-security prison, Ray Brook, near Lake Placid in the Adirondacks in upstate New York. Ray Brook had one thousand inmates and a comparatively relaxed, camp-like feel. Indeed, the facility was originally built as the athletes' village for the 1980 Winter Olympic Games and was comprised of a series of sand-colored residential cottages. The atmosphere was healthful and strangely rehabilitative compared with what Outlaw was used to. The mountain air was fresh and clean. At Ray Brook, he undertook business education and graphic design courses, and completed a rigorous 250-hour program to become a certified addiction counselor facilitated by St. Joseph's Rehabilitation Center in nearby Saranac Lake. The social worker wrote in her report that Outlaw's participation in the program was exemplary. "It was a pleasure to have Mr. Outlaw in class . . . he is a good student who is willing to learn."

In late August 2000, the Connecticut Department of Correction wrote a note to the Ray Brook warden stating that officers would be arriving between eleven o'clock and noon to transfer Outlaw to MacDougall-Walker, a prison in Suffield, Connecticut. No one at Ray Brook, however, told Outlaw of this development. Outlaw was playing basketball in the gym when he heard his name called over the intercom system. He didn't immediately respond because he was caught up in the action of the pickup game, but soon enough, a corrections officer appeared court-side and told Outlaw to go to the reception area, where he was greeted by four officers. Outlaw didn't understand what was happening until he noticed that the officers had Connecticut state flag patches on their sleeves. The glorious day had finally come: he was going home. "It's your lucky day," said one of the officers.

Outlaw was allowed to pack up his bags, then he was put in leg and arm chains, and placed in the back of a Chevy Suburban. Two officers sat in the front, and two in the backseat on either side of Outlaw. The Suburban drove through the high peaks of the Adirondacks, where deer were grazing and the early fall leaves were beginning to turn orange and red, then through Albany and down through the Hudson Valley, before cutting east to Massachusetts and turning south at Springfield. Six miles south of Springfield on Interstate 91, Outlaw reentered the state of Connecticut for the first time in eight years.

Upon arrival at MacDougall-Walker, which was just a few miles from Somers in northern Connecticut, the warden asked Outlaw what his intentions were. "I just want to finish my time and go home, sir," Outlaw said. Outlaw's release date, based on his reduced thirty-year sentence, was now supposed to be somewhere around the year 2014. The warden video-taped the interview, for reasons that were never explained to Outlaw. The next morning Pearl made the hour-and-fifteen-minute trek from New Haven in her new car, a gold Chrysler. In the visiting room, mother and son tearfully held each other for many minutes before they could speak. Holding her for the first time in almost a decade, Outlaw saw that she was grayer and had gained weight, but she was still his loving mother. "Please forgive me," Outlaw said. Pearl said nothing, but nodded.

Outlaw's stay at MacDougall-Walker was brief. Only a month later he was returned back to Somers, the place where he had started his prison journey, but which now had a different name—Osborne—perhaps in an effort to overturn its notorious reputation. Walking into his old prison, Outlaw couldn't help but think about the last time he was there, the night of the Interstate Compact transfer, when he was placed into that massive orange jumpsuit and had his wedding ring clipped off. The Feds never had paid for that ring, as they said they would. Outlaw felt like a different man. As he went to the yard and chow and was placed on the unit, Outlaw found that he didn't know many of the prisoners. A whole new generation populated the place, and Outlaw, now thirty-three, suddenly felt old. Then a veteran inmate recognized him. "Yo, O.G., how are you?" O.G. meant "original gangster," a sign of respect, and the other inmates took note. But Outlaw didn't go back to his former role of "original gangster." In the coming weeks and months, which then turned into years, he exercised in the gym, worked at a series of jobs around the facility (in the kitchen, as a hall monitor, in the library) and mainly kept to himself. He reserved his energy for visits with his children, who came when their mothers could bring them. The kids, now teenagers, were understandably reserved around him. They were glad their dad was home, but they didn't really know him. In the visiting room, he made them laugh, and he constantly apologized to them for not being there when they were young. Some old friends visited but not many. Many of the Jungle Boys were in prison, or had left the game, or were dead. Camile Leslie visited and asked for a divorce. Outlaw said yes, realizing he could no longer hold her captive. Camile had given birth to a daughter with another man during the time that Outlaw was away. He apologized to her, saying in effect that he had been out of his mind over the last ten-plus years with prison-inflicted trauma, but that his actions were his own. No one else was responsible.

When he arrived back in Connecticut, Outlaw was still classified as part of a "security risk group"—that is, gang-affiliated—based on his earlier tenure. But in his first month he told the director of security at Osborne that he wanted to formally renounce all ties with gangs within the prison system. In response, he received a letter on October 23, 2000, stating that based on his vow, his "security risk group has been placed on inactive status." And in fact, there was hardly a gang problem in the prison now. The free-for-all Wild West days of Somers of the late eighties and early nineties, which Outlaw had helped foment, were long gone. The anti-gang efforts of Mike Lajoie and his colleagues had been so successful that they had become a national model. Based on this work, Lajoie was now director of security for the entire Connecticut prison system.

Outlaw acquired one inconsequential disciplinary ticket in his remaining time in Somers. (The ticket was issued when Outlaw was in a group of eight men in the yard, one of whom taunted a female officer. Because the administration couldn't determine which inmate exactly did the taunting, everyone in the group was disciplined.) Based on his comportment, he earned significant "good time" credits: that is, reductions in his sentence based on compliance and positive behavior. In a series of hearings, his discharge date was moved up by four years, to somewhere in 2009 or 2010, depending, of course, on continued good "institutional adjustment." In 2007, he was transferred to a low-security prison, in Cheshire, only a tantalizing fifteen miles from New Haven. It was almost like he was home.

In February 2008, Outlaw attended a lecture by a man named Warren Kimbro, in honor of Black History Month. As a young man, Kimbro told the group, he had grown up in a middle-class family in New Haven and been a Korean War veteran. Rootless after the war, he became radicalized in the fervor of the late 1960s and joined the Black Panther movement. In 1969, he killed a man who he believed was a government informant. Kimbro's trial in 1970, and those of other Black Panthers, incited firebombing on the Yale campus and riots in New Haven, requiring the National Guard to be brought in. After being sentenced, Kimbro became a model prisoner, receiving a bachelor's degree in prison from Eastern Connecticut State University. In the forgiving spirit of the 1970s, Kimbro received a highly reduced minimum sentence—he only served four years in prison—and went on to receive a master's degree at Harvard, where he studied moral development with Lawrence Kohlberg, the leading international theorist in the field. Back in New Haven, he took over a struggling agency called Project M.O.R.E., which served ex-offenders in halfway houses. Speaking to Outlaw and others in the Cheshire prison, Kimbro was very open about the crime that he had committed, and said he regretted his action every day, every moment, and would for the rest of his life. He said he prayed for forgiveness every morning. Outlaw was drawn to Kimbro and could feel his authenticity. Kimbro's regret about his life oozed from every word he uttered. Outlaw was also impressed by Kimbro's style: he had distinguished salt-and-pepper hair and wore an impeccable pinstripe suit. Outlaw approached Kimbro right after the speech. "How can I get placed in your halfway house when I'm released? I'm supposed to get out in May." Kimbro advised him to write a letter to the parole board. "I'll try to make sure it happens," Kimbro said.

William Outlaw was indeed released to one of Warren Kimbro's halfway houses on May 21, 2008, with the plan of serving the last year of his prison sentence "in the community" at the facility. He had been incarcerated since October 1988, almost exactly twenty years. In that time, by conservative estimates Outlaw had cost the taxpayers of the State of Connecticut, as well as those of the United States, $2 to $3 million, an approximation, but certainly less than the amount of money he made while running the Jungle Boys. During Outlaw's time in prison, the incarcerated population in the United States had more than doubled, and the money spent on federal corrections had almost quadrupled. But with the exception of Richard Whitmire and Mike Lajoie, no officials among the thousands he had encountered within the prison system had inspired or "rehabilitated" Outlaw. No, the agents of change for William Outlaw were Frank James, a fellow prisoner; Marquetta, his daughter; and the deaths of two inmates, his friends. If Outlaw was rehabilitated, he was rehabilitated by his family and his friends—and by himself.

CHAPTER 7

Return

Quinnipiac Terrace, the housing project where Outlaw grew up.

THE LONG NIGHTMARE WAS OVER.

At six on a late-spring morning in 2008, Outlaw left his prison cell for the final time. As he walked down the corridors to the front gates of the prison, the guards were cordial. They didn't say what he had heard them say many times to departing prisoners: "See you later," or "We'll keep the lights on for you." They simply said, with what appeared to be sincerity, "Good luck." Before he left the compound, however, they put him in leg and arm shackles one last time. He was transferred to the custody of a Department of Correction parole officer who was going to drive him to New Haven.

When William Outlaw finally exited the prison's doors the first thing that he saw was a maple tree twenty feet away. Its leaves were fresh and clean in the sun, and the grass below the tree was saturated with a heavy dew. As he stepped forward tentatively onto the grounds outside of a prison for the first time in decades, Outlaw could not quite fathom what was actually happening. By this time he had been incarcerated for more than half his life. His Jungle Boys career, by contrast, had only lasted three and a half years. The fact that he was outside a prison was in itself astonishing: had it not been for Ira Grudberg's appeal, he would be in Cheshire or some other prison for another sixty years. He was leaving many people behind who had committed the same crimes as he had and who would not be released. That he was looking at a tree, and could walk up to that tree and touch its bark, was staggering.

"I'm going to give you a ride to New Haven today, Mr. Outlaw," the parole officer told him. "You'll find that a lot has changed." Outlaw clambered into the backseat of the van, smiling. It was a smile that wouldn't disappear for a week. But behind his joy, Outlaw knew the sobering numbers as well as any criminologist: of the approximately 700,000 people released every year from American prisons and jails, about half were back behind bars within three years. The highest risk was in the first year: almost 40 percent of released offenders were rearrested in the first twelve months. But Outlaw didn't know the even more urgent findings: that within two weeks of release, the risk of death—from drug overdose, cardiac arrest, homicide, and suicide—among released prisoners was thirteen times higher than that of the general public.

Outlaw felt confident about the changes he had made, but he hadn't proved anything to anyone yet. Many people, he knew, were expecting him to fail. He had known many former prisoners who had cracked up after they got out. Some had committed crimes just to be sent back, preferring the world they did know—"three hots and a cot," as it was often called, where there was at least a structure, no matter how mercenary and despotic that structure was—to an outside world that was now alien. As the officer drove the van into suburbia, Outlaw saw sprawling new big-box stores—Target, Dick's Sporting Goods, Bed Bath & Beyond, each half the size of a city block. He had never seen stand-alone stores of such dimensions before, nor did he recognize the names of the retailers. What Outlaw remembered as empty fields were now entire neighborhoods of condominiums. The cars were smaller than he remembered—some half the size of the big American cars he used to drive as a Jungle Boy—and the traffic seemed far heavier than it used to be.

THE VAN PASSED through the swampy area north of New Haven. Outlaw had always found this section around the highway depressing, as if a piece of the New Jersey Meadowlands had somehow been transported to the middle of Connecticut. It was almost reassuring that this area was absolutely unchanged, just as ugly and toxic-looking as it had been in 1988. As they approached the city, Outlaw could make out the small collection of skyscrapers that comprised downtown, and he felt a snap of anxiety in his chest. Lots of things would be waiting for him in New Haven, he knew. The whole town would be waiting to see if he had survived. A lot of people would be expecting, perhaps hoping, that he would fail. And some of the people he'd victimized, or their families, might be planning retaliation. Others, he knew, would want him to start up the gang again, looking to get on the gravy train.

Once downtown, the van merged onto Interstate 95, hugging Long Island Sound. Outlaw saw the spires of Yale rise amid the cityscape and he could make out passing glimpses of the harbor between buildings, glints of sun coming off the still-slate-gray water. How long had he dreamt of this moment! *Home at last.* Three blocks away, hidden from view by the monolith of Union Station, was the Jungle. Who was running the Jungle now? Surely someone was; drugs and crime, Outlaw knew, would infiltrate that place until the day they knocked it down.

The parole officer, who had been chatting amiably all the way, became silent as the van exited the highway, as if realizing the gravity of this moment for Outlaw. Outlaw too did not speak. The van pulled up on the sidewalk in front of the Walter Brooks halfway house, a three-story brick building that housed seventy men. The bricks were painted a soft pink, an unusual choice in predominantly gray New Haven. Other than its color, there was nothing distinctive about the facility. A passerby would not know that it was anything but an ordinary apartment building. The parole officer unlocked Outlaw's chains right on the sidewalk as surprised pedestrians walked past. He was greeted by Al, a six-foot-four soft-spoken man in his sixties who was a senior case manager, assigned by Warren Kimbro to supervise Outlaw. Inside his office, Al reviewed the usual reams of bureaucracy and informed Outlaw what he already knew: that he was still considered a prisoner—even if he was no longer in prison—serving out the end of his sentence, which would last one more year, in the community. His primary objective at the halfway house, Al said, was to get a job, but it didn't need to happen right away. A period of adjustment was to be expected; often released prisoners were in too much of a rush and became stressed, leading to relapses of various kinds. Outlaw was to submit to a biweekly urine drug test and attend various in-house therapeutic groups. The next two weeks were going to be an orientation period, during which time he would be prohibited from leaving the facility. Outlaw agreed to everything, but he couldn't really focus on what Al was saying. His mind was dancing with excitement and elation. Warren Kimbro, whose office was in another Project M.O.R.E. program building across town, stopped in just to say hello. "Welcome home, Juneboy," Kimbro said as he hugged him.

Al showed Outlaw his bedroom, a four-man room on the first floor, with two individual beds and a bunk bed. They then toured the facility—staff offices, a large kitchen, a rec room, and many dozens of bedrooms—where Outlaw met other residents and staff. Everyone knew who he was. Most of them called him Juneboy, even though they'd never met him. Apparently the legend of the Jungle Boys lived on. Later that day, Outlaw met with Kim Goodman, a petite blond parole officer who specialized in tough cases. Kim had previously read Outlaw's file and wondered what she was getting into, but she was experienced enough to know to look beyond the paperwork. Indeed, when she met Outlaw, she saw a likable and motivated person. "I like to keep things real," Kim said to him. "Given your size and reputation, there's already a target on your back. But I'm going to be here for you. There are only two things I expect from you: to respect me, and to tell me what is going on with you honestly, the good and the bad." Outlaw was struck by how similar her rules were to those of Richard Whitmire in Lewisburg.

Pearl visited on the second day. As they hugged, Outlaw's mind flashed to the conversation they'd had when she visited him on his second day at Somers all those years ago, when she'd said she could finally sleep once again now that he was incarcerated. Outlaw wondered if she no longer felt that sense of security now that he was out. Pearl had gotten more religious as she'd gotten older, and she talked constantly about God, and how thankful she was to him that Outlaw was out of prison. A few days later, Pearl made one of her famous soul-food dinners in the Walter Brooks kitchen for all the residents. She cooked for hours and put up balloons all over the ground floor. Outlaw's children attended, and the feast went on for hours.

His children, now in their late teens and twenties, were relieved that their dad was no longer in a jumpsuit, but they often maintained a wary distance. Outlaw, ever the adapter, settled into a kind of routine in those first few days: biweekly meetings with Al and Parole Officer Goodman, visits with family, including aunts and uncles, and more time with his children. Several former Jungle Boys popped in. Pong was working as a maintenance man, and Alfalfa had a job in the stockroom at Target. Other members of the old crew were now installing sprinklers, selling cars, working construction, and doing maintenance at Yale. Those who had the most success were working in sales, a natural evolution, of course, from selling drugs on the street.

Walter Brooks was filled with young men in their early twenties and even late teens. Some of them had been in prison for just a few months, for stealing a car stereo or a single instance of selling marijuana. Outlaw was older even than most of the staff, many of whom were in their twenties and just out of college. The kids listened to rap artists he didn't know, and songs that didn't make any sense to him. Who was Jay-Z? Who was 50 Cent?

After one of the in-house group sessions on avoiding the triggers of substance abuse, a young kid called him over. His name was Jasper. He had a hardness to him and was covered in tattoos. Conspiratorially, he whispered to Outlaw, "You're going to get it all going again, right?" Outlaw said nothing.

"This is all like . . . um . . . a front, right? This good-guy thing you're doing?" Jasper asked. "You gonna get back in the game, right? Because if you do, I'd like to get in with you. You could teach me a lot."

"Right," Outlaw said. "I'm going to get back in the game."

Outlaw could only smile. What else was there to do? He wanted to explain to Jasper all that had happened to him; how he was simply ecstatic that the whole harrowing nightmare was now over, and how he wished it had never happened in the first place. He wanted to talk about Frank James, the killings at Lewisburg, the phone call from Marquetta, but how could he possibly explain? And even if he could explain his story, how could a guy like Jasper possibly understand?

FIVE DAYS LATER, Outlaw lay in bed. The mattress sagged in the middle from his girth, and his feet protruded from the end due to his height. It was raining outside, the spring rains of New England that could go on for days. Outlaw pressed the pillow down over his ears and tried to block out the world. He knew he was supposed to go to breakfast—instant coffee and stale waffles—but he just couldn't manage to do it. He had not been able to sleep for five days.

At first the lack of sleep was almost a good thing. Outlaw was so astonished to be out of prison that he believed if he shut his eyes the dream would be over and he'd wake up back in Lewisburg or Leavenworth. He was overstimulated by all the novel sounds and sights. It was actually hard to break away from the decades of routine that prison had instilled in him. Now he had to decide for himself when he went to his room at night and what he wanted to watch on television. It was overwhelming. As much as he tried, he still couldn't sleep, not even for half an hour. By the fifth day, he was seeing and hearing things that weren't there, or at least that he thought weren't there. The images in his head didn't even make sense. Visions. Molecules in the air. Or birds exploding in the middle of flight. It felt like a psychedelic drug experience.

Al knocked on the door to Outlaw's bedroom. "You got to come down for nine o'clock count, Mr. Outlaw."

"Yeah . . . s-sure," Outlaw said, embarrassed that he was still in bed. He had a hard time even forming words. "Be . . . be right there."

Outlaw tossed the pillow away and held his hands over his ears again. He hadn't heard the sound of traffic for twenty years. He was beginning to think enduring it would be impossible.

SOMEWHERE IN THE middle of the long nights of sleeplessness, which went on for five more days, Warren Kimbro asked to meet with Outlaw in Al's office. Outlaw couldn't believe how crisp and debonair Kimbro looked. That day he was wearing a charcoal-gray suit, a purple shirt, and an indigo tie. It was remarkable to Outlaw that anyone could look so composed and stylish. He could barely think straight.

Kimbro said, "I need to tell you that the FBI is following you. I got a call from their New Haven office. They could be wiretapping your phone calls and they'll definitely be following you when you go back out on the streets."

Outlaw asked if this was some kind of nightmarish joke. "No joke," Kimbro said. "When I was a Black Panther, I was under constant surveillance by the FBI. I don't joke about the Feds. Be careful out there."

The next morning, after Outlaw endured yet another sleepless night, Kimbro made another special trip to see him, explaining that the week's food was being delivered that day to Walter Brooks, in a tractor trailer. The dozens of heavy cartons had to be unloaded and placed in the kitchen and basement, and many items—carrots, meats, potatoes—needed prepping. Some old beds and mattresses had also recently become infected with bedbugs, a common scourge at the halfway house, and needed to be brought to the dumpster and replaced with new beds that were arriving later that day. Warren asked Outlaw if he could help out with the day's activities. Outlaw enthusiastically agreed. He worked all day, never stopping for a break, and Kimbro stayed around, checking in with him.

In bed that night, in his room with three other residents, Outlaw was exhausted from the physical exertion. Every muscle in his sprawling body hurt. For the first time in more than a week, he slept that night, nine hours straight through. The next morning he felt like a new man in a new world. Reborn, as Pearl might like to think of it. When he came downstairs and looked out the kitchen window, he saw that the sun was yellow gold and the sky was a clear and flawless blue. The day was excellent and fair. Six days after that first intoxicating night of sleep, Outlaw passed his orientation period. Now he was finally allowed out of the house, on "job search" to find employment. He had not walked freely on a city sidewalk in more than twenty years, not since he had given himself up, accompanied by Ira Grudberg, at the New Haven police station and been sent to the Bridgeport jail. He found himself walking the seven blocks from Walter Brooks to the New Haven Green. Space stretched out in front of him. He could see four, five, six blocks ahead as he walked. The sense of openness made him anxious. The widest expanse he'd experienced over the last two decades was a prison yard. Looking down Howard Avenue was daunting, like edging out to the end of a diving board a hundred feet above a shallow pool of water. As he crossed the street, a car almost hit him. "Fuck you!" the driver screamed. Outlaw wondered if he should just stop. *No. One foot after the other. And then another. Keep on walking.*

He found a bench and sat down. He had not actually intended to go to the Green—there was not a business on the Green, which is the central and most corporate business district of New Haven, that would consider hiring a just-released felon—but it was as if he had no control over his destination. His body—his feet—had just taken over and led him to where he now sat. He remained on the Green for three hours. He watched the city buses; he watched the people in suits rushing to banks and restaurants; he watched the homeless people. Outlaw looked at the pigeons and the discarded newspapers blowing in the wind. He saw how people cut their hair, and how different the styles were from those of the 1980s; he heard the music from passing cars, also completely foreign to him; he watched the traffic, and studied the amount of space that drivers gave themselves when driving behind the car ahead of them. It was like he was experiencing these things for the first time. Everything was different. The parts of New Haven that hadn't changed—the historic campus of Yale, the few surviving stores from the 1980s—didn't match his memory of them. It all seemed like a drug trip out of the movie he had loved so long ago, *Once Upon a Time in America*. His life—the rise and the fall of it—had turned out to be exactly like the film. The whole trajectory had transpired just the way somewhere deep inside he thought it might from that moment in 1983 when Butchy had pulled over on Lilac Street in his Porsche and asked him to deal for him. It had ended up in prison, just as he feared that it would. He felt like a visitor from Mars. Or Rip Van Winkle. Or a bag man, more like it, living on the New Haven Green.

Outlaw's reverie was interrupted by a man who approached him.

"I don't fucking believe it. Juneboy, you're out! Oh my God!" Outlaw peered into the man's haggard face. He appeared to be around forty, and looked like he had been in the game, in the street life, for a long time. Outlaw stared into his face for some time, but it triggered nothing.

"You don't remember me?"

Outlaw slowly shook his head. "I'm sorry," he said. "I really don't."

"You're shitting me. We used to hang out, back in the day. The Jungle Boys! There ain't been nothing like that before or since, man! You sure you don't remember me?" The man was offended. *That's what happens when you're away for twenty years,* Outlaw thought. *You can't remember the most basic things.*

The man—who said his name was Grady—gave him the news of the last two decades. So-and-so had died of cancer at age thirty, someone else had moved to Atlanta, another person had started working at a car dealership and was now running the place. It all bled together in Outlaw's mind, and he couldn't quite follow it. Then Grady started talking about the people who had died. There were so many that Outlaw's mind began to shut down.

As Grady went on and on about the dead, Outlaw's eyes gravitated toward the courthouse, just a hundred yards away, on the corner of the Green. He let himself absorb the image of the building, with its white columns, its stately power. He remembered walking in for his trial with such bravado, thinking there was no way he was ever going to be convicted. Showing off his gangster's gold teeth for all the jury to see. His shock at the verdict. How arrogant he'd been! How stupid! Sitting on the bench, looking at the courthouse, not even noticing that the man with the drug-worn face had slipped off, Outlaw thought about that time in his life when he believed at least for a time that he could get away with anything he wanted.

Later, as Outlaw signed himself back in at Walter Brooks, the desk clerk asked to see his paperwork—signed slips from potential employers proving that Outlaw had interviewed with them.

"How'd the job search go?" the clerk asked.

"I'm sorry," Outlaw said. "I couldn't do nothing today. I just sat on the Green. I didn't get no pussy or nothing. I just sat on the Green, looking around."

The clerk smiled. "Thanks for your honesty," he said. "That's all right. You can go on job search again tomorrow."

They were about the same age, Outlaw and the desk clerk. "Juneboy," the clerk said, "I've been working here for a while. Most of the guys that go through here go back. Half of them go back in six months. From what I've seen, the guys that make it here—the ones that don't go back to the big house—are the ones who just listen. They don't talk a lot. They just listen." He paused a second. "That's my advice to you. Just listen."

"Thank you," Outlaw said. "I'll try to do that."

Warren Kimbro popped in. He had his briefcase and coat, apparently leaving for the day. "Juneboy, how was your job search today? You didn't do anything the FBI wouldn't like, right?"

"No, sir," Outlaw said. "No, I did not."

IN THOSE EARLY months after release, Outlaw began to feel comfortable when he was inside Walter Brooks, especially under Kimbro's tutelage and Al's steady support. He became a leader, a kind of ancillary staff member. One time a disrespectful young kid called Al a "white cracker."

Outlaw made him apologize to Al's face. "Big Al's a good dude, don't say that to him!" But outside he was far less at ease. He found that he had to relearn just about everything. Entire sections of New Haven had been torn down. Old mom-and-pop laundries and cigar stores were now chains like Gap and CVS and the Apple Store. On his first opportunity for an extended outing from Walter Brooks, Outlaw walked the three and a half miles to Q View, only to find that the original housing project built in the 1940s in which he had grown up no longer existed. Outlaw encountered a man walking down the street who told him that the city had razed Q View five years earlier. The old Q View had become so overrun by drugs and gangs, the man explained to Outlaw, that it was considered beyond saving. Outlaw looked at the spot where Q View had once been and saw a small village of handsome and well-constructed condos that were well maintained by the residents and the city. Outlaw recognized none of the new residents and departed after a few minutes, full of dismay that his old neighborhood, however toxic it surely had been, existed now only in his memory. He walked away from the new antiseptic development to the spot by the river where he used to escape as a kid, to the bunker-like space under the highway, where he would elude his mother and father. He decided to give this place a name. This was now his "bunking out" spot. "Bunking out" is a prison term for when you tell the corrections officers you can't leave your cell that day. For the first time in twenty years, Outlaw lay down on the concrete and looked up at the highway shaking above him from the thundering traffic. Just like when he was a kid, he thought of some place far away—any place that was not New Haven. But his mind couldn't find any place to go to. When he got up, he realized that he no longer needed that faraway place; he no longer felt the urge to escape. He decided that he was home. New Haven always had been his home, he realized, and always would be.

Sometime during that first summer of his release, Outlaw walked to the mayor's office. "I'm a resident of New Haven and I'd like to meet with the mayor," he said to the startled receptionist. She explained that the mayor held weekly office hours for citizens and Outlaw should return then. Outlaw returned twice and sat in the waiting room outside the office, but was never called in to see the mayor, although others were. He had wanted to deliver a message that he was a reformed man and wouldn't bring any trouble to the town.

The staff at Walter Brooks introduced him to the Internet for the first time. In prison, he had occasionally been able to work on computers, but they had not been connected to the Web. Now Outlaw saw Google and YouTube for the first time and searched his name and found old articles in the *New Haven Register* and the *Hartford Courant* about the Jungle Boys. He was stunned by the way everything was all connected. After being given a cell phone by an old friend, he spent four consecutive nights reading the manual. He also watched a lot of television, including *The Wire* and *The Sopranos*, both of which he found entertaining but not credible.

WATCHING TELEVISION, HE kept on seeing a Dunkin' Donuts commercial for a new sandwich made of flatbread. He had no idea what flatbread was and resolved to go to Dunkin' Donuts and try it. One might think such choices were trivial, but Outlaw had learned to treasure such decisions. They had been denied him for twenty years. He went to the Dunkin' Donuts on Whalley Avenue, and ordered flatbread. He wasn't impressed by its taste. As he ate it, he spotted a Help Wanted sign. The store supervisor saw him looking at it.

"Looking for a job?" she asked.

"Not really," Outlaw said. He didn't think he was ready.

But she pressed. "Why not?" she said, and handed him a job application.

Outlaw sat down at a booth and filled out the form in his careful handwriting. He wrote in his name and address. Then came the challenging part: what Al and all the counselors at Walter Brooks called the "felony question." It was on every job application: "Have you ever committed a felony?" Al had advised him to answer this question honestly, and explain to prospective employers in simple terms what he had done, and then state, with conviction, how he was moving on with his life. This advice was fine in theory, but many employers would discontinue the interview the moment you acknowledged a criminal history. Outlaw skipped the question and went on to the section on professional and personal references. Here too Outlaw was stumped. He hadn't had a legit job in his life. Finally he wrote the name Mike Lajoie. To the query "Nature of your relationship," Outlaw thought for a moment, and then wrote the word "Friend."

When he was done with the form—except for the felony question—he went to the supervisor. "Oh, one thing I got to tell you, I just got out of prison after twenty years. I just wanted to tell you that personally, rather than write it down on the paper. But I'm here to tell you, I'm ready to turn my life around." She listened, and they talked about other things, including what the job entailed and what flatbread tasted like. By the end, the supervisor said, "I can't believe you were in prison for that long. You present yourself so well. If your references check out, you've got the job."

The supervisor called Lajoie, who spoke generously of Outlaw. But streetwise as always, Lajoie didn't exactly offer how he had first gotten to know Outlaw, nor how this particular Dunkin' Donuts applicant had once smuggled a gun into prison. Lajoie merely said Outlaw was "resourceful" and he had "excellent social skills." "You'll love him," Lajoie said. "He's a great guy."

Outlaw was assigned the four-to-midnight shift at the drive-in booth, where he dispensed coffee and doughnuts. At first the store couldn't find a uniform to fit him, so he simply wore a red T-shirt with a Dunkin' Donuts cap. The preternaturally calm Outlaw, who could easily stay composed during a shoot-out, became extremely nervous before each shift. He wasn't sure he could handle being exposed to the general public. He wondered if he could relate to people anymore. But he regained his confidence quickly. "Your coffee's not hot?" Outlaw would say, smiling. "It's on the house."

On his first nights in the booth, the store was busier than it had ever been. Everybody in town, it seemed, wanted to see where Juneboy had ended up. Outlaw was mainly delighted by the attention. It felt good to know he was still relevant. Actually more than relevant: he was still famous in town. The store owner, Abdul, was delighted, but also confused. After business kept on booming like this for days, he finally asked Outlaw, "Who the hell are you?" Outlaw just smiled back at him with his gold teeth.

DUNKIN' DONUTS WAS two and a half miles from Walter Brooks, and there were certainly times on his long walk to and from work that Outlaw thought about peeling off to Whalley Avenue and entering nearby Newhallville to return to the kinds of activities that he had left off two decades ago. Probably in an hour, he could have $500 in his pocket, far more than he made at Dunkin' Donuts in a week. He knew from the younger men at Walter Brooks that the gang scene in New Haven was much more of a ragtag affair than it had once been, comprised of small "pop-up" crews of five or seven guys, far different from the mini–corporate enterprises of the Jungle Boys and the Island Brothers. He had no doubt that he could whip a street gang into shape in a matter of days. The young kids would naturally defer to his seniority, expertise, and size, and he could probably make $30,000 or $50,000 in several weeks.

But he never made that detour. Over time, he grew to love his $10-an-hour job at Dunkin' Donuts, and considered it a gift from God that he'd been led there by that commercial. Outlaw did need to save up a lot of money to get an apartment on his own before he left Walter Brooks, so he picked up a second job at a nonprofit that did substance abuse counseling in Newhallville, helping out as a billing and office assistant, ordering supplies, and sweeping up the office. He left Walter Brooks at 6 A.M. and came back at 1 A.M. Al and Kim Goodman had never seen any of their parolees work so hard.

In addition, he enrolled in a class at Gateway Community College, a beautiful new facility in downtown New Haven, with the goal of getting his associate's degree in human services. He wanted to carry through on his vow to himself after completing his G.E.D. at Lewisburg that he would continue to pursue an education for himself. But the school administration wouldn't accept his credits in substance abuse counseling from Ray Brook, which the prison had told him would be transferrable to the outside. Without those credits, an associate's degree would take years of time and expense. Outlaw left angry and bitterly disappointed.

Outlaw always knew that he was at risk in choosing to return to New Haven, perhaps dire risk. At various times, he had considered moving to North Carolina to start fresh, where he could avoid the New England winters and where no one might be waiting to settle an old score. Outlaw was beginning to hear rumblings about old enemies of the Jungle Boys who were not thrilled that he was back in town, and in particular that he'd beaten an eighty-five-year bid when others who had worked against him and even for him were still locked up. In fact, one former rival had recently been picked up by the police and imprisoned on a weapons charge. Word on the street was that he'd been planning to use the gun to shoot Outlaw.

Outlaw had heard from the scuttlebutt around Walter Brooks that another man named J.R. had been talking about shooting him. In the late 1980s, J.R. had been a "stickup" guy who did random freelance robberies around town. After he'd attempted to rob the Jungle Boys, the crew had savagely beaten him, and apparently J.R. had never been able to let the humiliating event go. More than twenty years later, J.R. still wanted to get revenge on Outlaw.

One morning Outlaw got out of bed at five, walked to the train station, and got a cab. Fifteen minutes later he stood in the hallway of a burned-out apartment building. Outlaw knocked on the metal apartment door and J.R. opened it. Despite appearing terribly hungover, J.R. couldn't disguise his apparent shock at seeing Outlaw, whom he'd last encountered twenty-odd years earlier.

Outlaw didn't give J.R. any time to react. "Look, man, we got to talk," Outlaw said. "I came here to talk to you man-to-man, to show you that respect, to give that respect back to you. I need to tell you that. But check this out, man: I heard you were talking reckless about me since I been back in town. I need to tell you what's up. That shit you been saying about me ain't sitting well with me. I ain't got no beef with you. I trust you. But you gotta understand, we told you, I warned you personally, not to come in the Jungle, back in the day. We said we didn't tolerate no stickup boys. It was fair warning."

J.R. said, "But your guys destroyed me, man. I never had the same rep around town after that."

"Yeah, you right," Outlaw said. "I'm not going to say that didn't happen. But I personally came out here at this hour to tell you I respect you highly. I ain't got no beef no more. That was a long time ago. I'll give you the benefit of the doubt that nothing more is going to come of this."

J.R., still flabbergasted from the encounter, nodded.

Outlaw said, "I'm going to go to work now, and I'm going to go in peace knowing that this shit is behind us."

As Outlaw excelled at Walter Brooks, it became clear that Warren Kimbro was grooming him for some kind of leadership role in his organization. Kimbro most likely saw a lot of his young self in Outlaw. One day he came to Outlaw and said, "We're going to get you a real suit so I can take you out on the road." He brought him to his tailor, who made a specially designed dark suit just for Outlaw. "Now, we're going to get you some opportunities to tell your story," Kimbro said. In those last few months of his time at Walter Brooks, in addition to his two jobs, Outlaw began to accompany Warren Kimbro to meetings and speeches at schools, civic organizations, and advocacy groups around the state. Kimbro, having turned Project M.O.R.E. into a well-run and growing agency, was now highly respected in the community.

On February 2, 2009, Kimbro and Outlaw spoke to a group of juvenile offenders in Bridgeport. Kimbro dropped Outlaw off outside Walter Brooks House. "See you soon," Kimbro said to Outlaw, as Outlaw exited the car into the pouring rain. A few hours later, Kimbro complained of chest pains and was rushed to Yale New Haven Hospital. He died of a heart attack that night. Outlaw thought it was a bad dream when he was awakened in the middle of the night with the ghastly news. He was given special permission to go to Kimbro's house, with the veteran members of the staff, to console Kimbro's son and widow.

The obituary in the *New York Times* read, "Warren Kimbro, Ex-Panther Who Turned to Life of Service After Killing, Dies at 74." The article quoted Kimbro from an interview in 1973 in which he had reflected on his murder of Black Panther Alex Rackley in 1969, and his rehabilitation. "I was just a kid out there who didn't know how to handle himself, and it was a slap in the face with cold, hard reality that turned me around back to what I was. . . . I'd lie awake in my cell at night trying to figure out what makes me tick, and I succeeded. I'm now what I was before 1969."

Outlaw was absolutely devastated by the death of his mentor, and served as a pallbearer at the funeral. He cried so heavily during the service that he soaked his shirt.

He only had a short time with Kimbro, but to Outlaw he'd lost the father that he'd never had.

ONE MONTH AFTER Kimbro's death, Mike Lajoie, who had heard about Outlaw's speaking prowess, wrote this note to the warden of a Connecticut prison:

> On Tuesday, March 31, 2009, 1:00 pm, I am giving a motivational presentation to a group of approximately 50 to 60 inmates at the Corrigan-Ragowski Correctional Center. The topic is "Preparing for Discharge, Being Successful." I would like to request that Mr. Outlaw be permitted to attend and assist in this presentation. . . . Thank you for your consideration of this request. Michael Lajoie, Director of Security.

Kim Goodman, Outlaw's parole officer, would have driven Outlaw to the event but he didn't fit in the back of her squad car. Outlaw was driven to the prison in a Walter Brooks van, where he was met by Kim. Mike

Lajoie greeted him in the lobby. Lajoie was now in a suit and tie rather than a corrections officer's uniform. To Outlaw, Lajoie hadn't aged much. Yes, his forehead was a little higher, the lines of his face etched a little deeper, but he still looked young and vital, and even more muscular than he had been. To Lajoie, Outlaw too looked essentially the same. The two men hugged. It was a joyous reunion between the old purported enemies. "I'm so proud of you, Juneboy," Lajoie said.

"Thanks, Lajoie. I learned from you, man. You were the real deal, even when we were butting heads," Outlaw said.

Lajoie and Outlaw walked into the auditorium, where one hundred inmates were waiting. Lajoie spoke about his long career, and how he had worked with some of the most hard-core criminals. He said that Outlaw was an example of how it was possible for even the "worst guys" to change. Outlaw talked about his prison odyssey, and how federal prison had "kicked his ass" and humbled him. He introduced Officer Goodman to the men, calling her his "warden in the community." He said that "if you get the right parole officer when you leave here, they are there to help and support you. They want to keep you from going back to prison. Otherwise they don't have a job!" The crowd laughed. Afterward the men flocked to both Lajoie and Outlaw. Lajoie said privately to Outlaw, "You can do this. You can really speak. You should do this for a living." Kim Goodman teared up.

Not long afterward the FBI in New Haven called Mike Lajoie. In his position as director of security for the Department of Correction, Lajoie had known the FBI were following Outlaw upon his release. They said they'd been tracking Outlaw for months now, and he had done nothing even remotely suspicious. The FBI officers were stunned. "Guys like Outlaw always get back in the game," an agent said. Lajoie said, "I told you he wouldn't. I just knew it. Outlaw is an old-school gangster. His word is his bond."

THE END OF Outlaw's sentence was soon approaching. On June 1, 2009, he would be released from all obligations to the State of Connecticut. At

Outlaw's discharge from Walter Brooks, Al told him that he was the best client he'd ever had, and that his presence had contributed to a particularly quiet and peaceful time in the life of the house. When June 1 came, Outlaw met with Parole Officer Goodman for the last time. She gave him an official letter indicating he was done with his obligations to the state. Outlaw gave Kim a bear hug. Kim said he had "done great," as well as anyone had done in her time as a parole officer.

Outlaw moved to West Haven, five miles north of New Haven. Through a church friend of Pearl's, he got a discounted rent on a modern, nicely appointed apartment in a complex adjoining the Target parking lot. "For you, given your situation, I won't raise the rent," the landlord said. Outlaw had saved up $7,000 while at Walter Brooks, but he also had some extraordinary benefits to which few returning offenders have access. A former drug dealer in the Bronx with whom Outlaw had become friends in the 1980s was now a legitimate entrepreneur, owning a series of bodegas. He gave Outlaw a used white GMC SUV, perfectly matched to his size and style. Other former dealers—or perhaps current dealers—gave Outlaw tens of thousands of dollars as a welcome-home present. In turn, motivated equally by guilt and generosity, Outlaw gave most of the money to his kids.

In his new place, Outlaw enjoyed a bachelor life. Plenty of women, still drawn by Outlaw's reputation and charm, came around to see him. He began making up for twenty-five years of lost time. He had a short-lived relationship with one woman, whom he met in the checkout line at Walgreens. Outlaw was not careful with birth control, and fathered a boy, whom they named Joshua. Outlaw and Joshua's mother soon broke up, but Outlaw provided financial support and swore he'd be there for Joshua in a way that he never had for his other children when they were young.

Outlaw continued to love his work at Dunkin' Donuts. It had given him confidence that he could reenter the workforce. But he wanted to follow through on the plan he had created with Richard Whitmire. He had heard about a new program in town, sponsored by the City of New Haven and run by a nonprofit. On the face of it, the new project seemed like a crazy idea. It involved the hiring of former offenders to go into the stressed neighborhoods of New Haven and work with at-risk kids and advise these young charges, from the bitter experiences of their own lives, to stay clear of crime.

To Outlaw, it was an incredible concept. Maybe his criminal career, which had defined him most of his life, could become an asset, even a job qualification. They should be interested in hiring him. Because, after all, who in New Haven knew more about crime than William Outlaw?

CHAPTER 8
The Interrupter

Outreach on the streets.

THE ROOTS OF NEW HAVEN'S STREET OUTREACH WORKER PROGRAM went back to the summer of 2006, when the city had become so appalled by the murder of young people that it was willing to consider anything, even radical ideas, to stem the violence. Two particular events prompted the desperation.

At a few minutes before eleven on a June night in 2006, Jajuana Cole, a thirteen-year-old girl, was in her apartment in Dixwell, a few blocks from Yale. Jajuana was with her friend Krystal, when the two girls heard the sounds of a block party down the street. Curious, they walked down the stairwell of Jajuana's building, past graffiti that read

FUCK TRIBE (a reference to an ongoing feud between the 'Ville gang from Newhallville and the Tribe gang from Dixwell), and walked one block to where a group of young people were listening to music and hanging out on the warm summer night. Some of the young men were members of the Tribe gang. A few minutes later, five teenage boys, all members of 'Ville, drove through Dixwell and spotted the Tribe members at the block party. The 'Ville boys parked their car, and two of them got out, armed with guns, and snuck through a backyard. They were followed by a third boy who videotaped the entire proceedings.

Krystal saw the boys coming. "Please don't shoot!" she screamed.

"Shut the fuck up!" one of the 'Ville boys responded, and opened fire.

On the videotape, which was shown later in court, a nineteen-year-old man was seen to fire bullets from a .380 automatic into the crowd. The video showed the girls screaming as the bullets flew. Jajuana was shot in the back, presumably in the act of fleeing, and killed, and Krystal was wounded but survived.

The next day, members of the community congregated on the street by Jajuana's home. Friends and neighbors remembered the young girl by her nicknames "Nonnie" and "Pretty Toes," the latter reflecting her love of dance. A minister declared, "Let us sound a trumpet to all parents who know their children are in possession of a handgun—we will not tolerate it. If your child or children are possessing a handgun, we want them arrested."

Days later, hundreds of people, including many of Jajuana's classmates, attended her three-hour funeral service. At the pulpit New Haven's mayor John DeStefano said, "There are more victims than just this child . . . all the children that had to see this," he said, addressing the crying children in the pews. "The question is: 'After today, what do we do?'" Twelve years previously, DeStefano, then newly elected, had spoken from this same pulpit at the funeral of Danielle Monique Taft, who had died when she was seven months old. A young man sprayed fourteen bullets from a stolen 9 mm semiautomatic into the window of a ground-floor apartment, instantly killing Danielle and permanently paralyzing her grandmother. At the end of the service, Jajuana was remembered as a leader on the local drill team. The squad leader, forty-year-old Doug Bethea, addressed the congregation. He said that Jajuana—"Pretty Toes"—was a lovely girl with a lovely smile. The drill team marched and exploded into song. "We love you, Nonnie!" they belted out.

One month later, Justus Suggs, also thirteen, was riding his bike in the Hill when he too was shot in the back. Justus Suggs did not know—nor would he ever know—that earlier that night a boy had been beaten by another boy from the Hill in a fistfight downtown. The humiliated boy went home, got a gun, was driven by a friend to the Hill, and fired the gun indiscriminately into a group of teenagers, including Justus. Some escaped the spray of bullets, but Justus Suggs did not. He was brought to Yale New Haven Hospital, where he went into a coma. Suggs died thirteen days later when his mother took him off life support.

Jajuana Cole and Justus Suggs were just two of the murdered residents of New Haven in 2006. There were twenty-two others.

The following year, in 2007, the police department's chief of narcotics and two detectives pled guilty to federal charges of taking bribes from bail bondsmen and stealing cash from crime scenes. Under pressure from these scandals and the rising homicide rate, the police chief resigned. Just prior to his resignation, a private consulting firm, hired by the mayor and city council, issued a 133-page report on the state of the police department. The report recommended a major overhaul "in order to regain the public trust." The report went on to say, "Officers seem to have little respect for citizens based on how community members report they are treated and spoken to, especially the youth in the community. There is a racial divide within the department and this internal division is reflected in an external division between the department and the community."

In the outrage that followed the deaths of Jajuana Cole and Justus Suggs, there was talk of imposing a 9 P.M. citywide curfew on youth. That notion was rejected, but many civic leaders agreed that in order to avoid a return to the unprecedented violence of the 1980s—the era of the Jungle Boys—the city needed to try something entirely new.

In the previous decade, a number of high-violence cities around the country, including New Orleans, Chicago, and Baltimore, had created "street outreach teams" comprised of former offenders, presumably now reformed, to use their experience as cautionary tales to dissuade youth from entering gang life. The early results had been promising. Shafiq Abdussabur, a veteran police officer and activist in the African-American community, was the first person to raise the idea of bringing such a team to New Haven. The closest such team geographically was in Providence, Rhode Island, and run by the Institute for Nonviolence. The institute's mission was based on the work of Martin Luther King: to teach "by word and example the principles of non-violence and to foster a community that addresses potentially violent situations with non-violent solutions." The newly formed Providence team appeared to have had a remarkable impact: in 2005 and 2006 there had been 22 murders each year in Providence; in 2007, there had been 2. Abdussabur, along with other police and city officials, engaged in a series of meetings with their counterparts in Providence, and the Providence team offered to train a New Haven team should it be funded.

The funding for a New Haven team came together surprisingly quickly. Private sources—including Yale, the United Way, and the local community foundation—contributed $400,000. An additional $200,000 came from the state after the mayor and chief of police called the governor. A competitive bid was issued by the city, which was won by New Haven Family Alliance, a nonprofit agency that had an existing mentoring program for at-risk kids. "We already know many of these families," said Executive Director Barbara Tinney, an African-American woman in her late fifties who had grown up in the neighborhoods she now served, and earned a master's degree in social work from Smith College.

The New Haven Street Outreach Worker Program was officially kicked off with a press conference at city hall in July 2007, one year after Jajuana Cole and Justus Suggs were killed. Many people received the new project with dismay. Many officers, and members of the public, were openly hostile to the idea of hiring former felons, regarding the concept as the law enforcement equivalent of having the inmates run the asylum. The following comment in a newspaper article about the outreach team was typical:

Unsnuff the police and let them do their job! No more tax-payer funded pro-grams for these gun slinging-drug dealing-animals! The city is as violent as it ever was. LOCK'EM UP!

Doug Bethea, Jajuana Cole's drill instructor, was the first person hired on the team. Bethea, six feet tall and utterly lean—his cheeks just about sink into his face—was imbued with an infectious high energy, and would become the only person on the new team who did not have a felony. In fact he had been an assistant investigator in the police department for a decade, and continued to work as an emergency medical technician. Perhaps to add a measure of "street cred," he often went by the nickname "Doug-E Fresh."

Doug came to the outreach team along a route different from that of other team members. At 4 P.M. on November 27, 2006, just months after Jajuana Cole was killed, Doug saw his oldest son, Scotty, alive for the last time. Scotty had just turned twenty, and Doug gave him a gold chain for the occasion. Scotty thanked his dad, and said he was going to say a quick hello to his aunt Michele, who lived a few blocks away. Fifteen minutes later, Doug received a call from Michele that Scotty had been shot and was lying on the front steps of her townhouse, bleeding and unable to speak. Doug arrived in time to see his son being lifted into an ambulance. Doug jumped in. At the beginning of the ride to the hospital, Scotty's hand was moving—quivering in shock—but by the end of the trip Scotty's hand was still. He was pronounced dead on arrival. The next day, Doug suspected that the motive was robbery: the gold chain that Doug had given him was gone. The murder was not solved until years later. Bethea said that he joined the street outreach team because "I don't want anyone else going through what I went through."

To start the project, the police gave the team a list of 200 young people, male and female, ages 13 to 21, whom the department considered at risk for, or already involved in, violent behavior. Outfitted in purple jackets, Bethea and the team walked the streets of the Hill, Dixwell, and Newhallville attempting to build relationships with kids and their families. In the first four months, the team engaged one hundred more kids than expected. A few months into the team's existence, a dozen Dixwell boys were shooting baskets with members of the outreach team when a car with Newhallville boys pulled up. Even after the Jajuana Cole murder, Dixwell and Newhallville continued to war with each other. A Newhallville boy pointed a gun at the basketball court. The outreach workers told the Dixwell boys to scatter, and then intrepidly approached the car and said to the boy brandishing the gun, "We need to talk before someone gets hurt." Later that night, the outreach team brought the leaders of both groups together for a tense meeting at "the Mudhole," an open-air lot. Under Bethea's guidance, the two sides took turns sharing their grievances. The outreach workers kept the tone as neutral as they could, and inserted multiple references to the recent murder of a seventeen-year-old, a suspect in three homicides himself. In subsequent days the outreach workers continued to meet with the leaders of both groups, and eventually brought them together for a meeting in which, remarkably, both crews signed a document pledging not to shoot each other. The team followed up with job-training sessions and sponsored a series of Newhallville and Dixwell basketball tournaments, which transpired peaceably. In quick succession, the outreach workers negotiated three more truces: between Newhallville and the Tre, Dixwell and the Tre, and Dixwell and the Hill. From city hall on down, everyone agreed: nothing like this had ever happened in New Haven. Even the more skeptical members of the police force were impressed.

Philosophically, the outreach team was influenced by the "Cure Violence" health model, originally developed by Dr. Gary Slutkin of the University of Illinois at Chicago School of Public Health, which amounts to a fundamental reconceptualization of youth violence from strictly a criminal justice problem to a public health problem. The model identifies three principles employed to combat epidemics of disease and applies them to youth violence: (1) interrupt transmission of the disease; (2) reduce the risk of the highest-risk cases; and (3) change community norms. In the case of combating youth violence, the approach translates to interrupting potentially violent conflicts after a shooting; targeting the young people who are known to have been behind the violence, who could be conceived in this model as spreaders of the disease; and changing the local culture by engaging the community at all levels through such activities as distributing engagement materials, holding community events, modeling positive behavior, and generally sending a message that violence will not be tolerated. Evaluations by the Centers for Disease Control and the National Institute of Justice on the effectiveness of outreach teams showed that killings were reduced by 56 percent in one neighborhood in Baltimore, and shootings were reduced by 41 to 73 percent in seven communities in Chicago.

Like many radical ideas, however, the idea of the outreach workers had ancient roots. The notion of deploying the very people who created the problem to generate its solution is a long tradition in Western culture (and other cultures), dating to Saint Paul on the road to Damascus and Saint Augustine. After all, sinners are the ultimate experts in the nature of sickness and wrongdoing. In his later writings, Carl Jung called the phenomenon "the wounded healer" and suggested that having a disease was the best training that a doctor could receive. "A good half of every treatment that probes at all deeply consists of the doctor's examining himself . . . it is his own hurt that gives a measure of his power to heal," Jung wrote.

William Outlaw, as he immersed himself once again in the community, began hearing more and more about the outreach team. As much as he loved his job at Dunkin' Donuts, he desperately wanted to join the new initiative, which was by then three years old. He knew many of the team members from his time as a Jungle Boy, even if a number of them had once been his sworn rivals. Outlaw certainly appeared to have all the requisite credentials to be a "wounded healer." Arguably, or perhaps inarguably, no one else on the streets of Connecticut had more street cred than Outlaw. No one in Connecticut had run a gang of the Jungle Boys' size and sophistication; no one else had been originally sentenced until the year 2073; no one else had been a shot caller of a hundred men in prison; and no one else had survived nearly a year in solitary in Leavenworth. And certainly no one else who even approximated those credentials was six foot four and four hundred pounds with gold teeth and a Mohawk. Outlaw set up an interview with Barbara Tinney. Tinney had heard the stories about Outlaw, but she chose to look beyond them, and was impressed by the respectful, motivated man she saw in front of her. Outlaw applied for the job and was immediately hired in 2010 for $15 an hour. For once, the "felony question" on the job application was not an issue.

The job description that Tinney handed Outlaw included the following activities: Be in the neighborhoods constantly, building strong relationships with youth, residents, businesses, and community groups. Form relationships in particular with the highest-risk kids, and the people surrounding them. Proactively intervene in circumstances in which gun violence is likely. When a shooting occurs, marshal an organized response to the shooting. Learn why the conflict happened in order to understand how to mediate and prevent violence in the future. Be omnipresent after a shooting and diffuse the desire for retaliation. Be willing to go to any neighborhoods where violence is an issue, even outside the city. Document all contacts and interventions. Engage in what are called "prosocial activities"—community events, basketball leagues, job training—any positive activity that keeps kids away from the streets.

Outlaw jumped into the work in his typically hands-on and ebullient fashion. He had a relentless, boundless professional energy that had been hemmed in after twenty years behind bars. Often he was the first to show up and the last to leave. In the early days, Tinney deployed Outlaw as a speaker to youth at high schools. He had developed a reputation as a fluent and moving public orator, and Tinney believed that the fact that he was so recently out of prison would give his message a particular resonance. After a speech at Hillhouse High School, Outlaw was cornered by seventeen-year-old Big Cooper. Big, at six foot three, three hundred pounds, was a decent student and a star offensive and defensive lineman on the football team. In a quiet area of the auditorium after Outlaw's speech, Big Cooper poured out his story. Football, he said, was his "angel," the sole stabilizing force of his life, so much so that he had a football with wings tattooed on his shoulder. Big lived with his mother and stepfather in Newhallville. Big was often forced out of the house after squabbles with his parents and he was frequently homeless, forced to crash with his "homies," who were involved in low-level drug dealing. During these times, Big too periodically dealt drugs and handled a gun. At school, he was often targeted by other boys because of his size. Big would never back down, and was frequently suspended for fistfights. Outlaw listened intently to Big, told him that he understood and related to his story, and the two exchanged numbers.

Big called Outlaw sometime later. He was upset. It was the middle of football season, and Big had gotten his grades up and was being visited weekly by college scouts. A student named Taquan, known to be gang-involved, had earlier that day threatened to shoot Big Cooper, and was telling others that he was going to follow through with the threat. The beef had started with a simple bump in the hallway. Taquan had collided with Big in the crowded corridor between classes, and given Big a foreboding look afterward. The next day, Taquan turned his back on Big in the cafeteria, and then turned around to wink at him seductively in front of other students. To Big this was a clear sign of "disrespect." When Taquan continued the taunting, Big, almost double Taquan's size, flew up from the cafeteria bench and attempted to body-slam him to the ground. When Big returned to school after the inevitable suspension, Taquan approached him and said, "Watch your back. I don't fight, I shoot."

In tears over the telephone, Big Cooper told Outlaw, "This kid don't have nothing to lose, but I do. I don't want this stupid beef to mess up my future." Outlaw said, "Don't worry about it. I'll take care of it." Outlaw, who knew Taquan, visited him later that night. He told Taquan that Big was a good kid, that Big didn't know his own strength when he nearly body-slammed him, and that he wasn't interested in any kind of ongoing beef with Taquan. "Man-to-man," Outlaw said to Taquan, "I am showing you a lot of respect by coming all the way out here to meet with you. Now I need you to lay off." Taquan didn't say anything, but the fact that the request was coming from Outlaw, whom many in his circle considered a New Haven legend, potentially gave him an "honorable exit" from the conflict. He could save face with his homies by saying that Outlaw had told him the conflict wasn't worth his time and energy. The threats to Big Cooper ended immediately. Big finished football season and graduated, and went on to play college football in Delaware.

Some of Big's friends were not so lucky. Radcliff Deroche, Big's childhood friend, was shot at three-thirty in the afternoon on Easter Sunday and died the next day. In the summer after Big's freshman year in college, his best friend, Daryl McIver, was shot and killed in the middle of the night in the Hill. After the second killing, Big went into a month-long shell, barely able to get out of bed. Big considered getting a gun to avenge his best friend's murder. But instead, Big Cooper called Outlaw, who picked up immediately. Later that day in Big's bedroom, Big told Outlaw that he thought he was suffering from depression. "That's okay," Outlaw said. "It's okay for a big strong guy like you to feel depressed. People think that tough folks like us don't have emotions. But I used to cry myself to sleep in prison sometimes." Outlaw set up a regular meeting schedule with Big, and later that summer, with other outreach workers, he took him and other boys to a college football game in the New Jersey Meadowlands. In time, Big added two more tattoos on his arm, just under the winged football: the dates of birth and death of Radcliff Deroche and Daryl McIver. "I don't know what I would have done without Outlaw," Big said. "I would have ended up like them probably."

As Outlaw grew into his role on the team, and experienced these kinds of successes, his work increasingly gravitated back to the streets. After all, he had always felt most comfortable on the streets of New Haven. He loved the ramshackle neighborhoods and dirty corners and busy bodegas and bars. He canvassed the neighborhoods nightly, house by house, block by block, individual by individual, and gang by gang. At first the younger kids didn't know who he was. But if they didn't know him, their uncles or aunts or parents did. Often there was a commotion when Outlaw appeared. "Juneboy's back!" older men and women would shout. "Oh my God! I can't believe it." The kids learned quickly that Outlaw had been a special kind of player. But of course, going back to the streets was also enormously dangerous for Outlaw. An old rival could step out from a car at any moment and blow him away. Outlaw constantly scanned alleyways and streets and cars for threats. Joining the street outreach team, in fact, may have been part of Outlaw's strategy for self-preservation: he may have thought it wise to create a high enough public profile so that anyone who harmed him, or killed him, would attract way too much heat from the authorities.

Within a few months of Outlaw's joining the team, a new beef emerged between another gang in Newhallville, called R2, and a crew on Derby Avenue in the nearby West River neighborhood. Three successive shootings, each clearly in retaliation for the other, occurred between the two gangs within a week's time. Barbara Tinney suggested that Outlaw try to engage with the kids in R2, and that Trent Butler, another outreach worker, approach the kids on Derby Avenue. Butler and Outlaw had been rivals in the 1980s, when Butler had led a small gang in Newhallville. In deploying Butler and Outlaw together, Tinney sought to demonstrate how it was possible for former enemies to now work side by side. After intensive lobbying, Outlaw and Butler convinced the leaders of the two gangs to come to the outreach team's offices. In the hallway outside the meeting, Outlaw patted the boys down to make sure they hadn't brought guns. They had not. Entering the conference room, the crews were greeted with a spread of pizza and chicken wings, which Outlaw had preordered. Outlaw had even bought hand sanitizer for the kids to use.

As part of his orientation to the team, Outlaw had received training on gang mediation, but more often he relied on his own insight and intuition gleaned from decades on the streets and in prison. Outlaw, typically so uproarious, spoke calmly and slowly, at times even in a whisper. First he let the kids simply vent. The kids were loud and aggressive to start, but after the initial tension eased just slightly, Outlaw announced the goal of the mediation: "We can settle this thing on our own," he said, "or we can do it with the help of the funeral home or the courthouse. It's your decision, know what I'm sayin'?" Outlaw also stated up front that his intention was not to stop the kids from dealing drugs—"If you gotta make money, do what you gotta do"—but rather to stop the violence. "You guys don't have to be friends, neither," Outlaw said. "You just gotta not hurt each other."

They asked the kids to articulate exactly what the beef between the two groups was. As the boys spoke, it became clear that the conflict in the neighborhoods dated back thirty years and had led to an endless tit-for-tat series of retaliations that were often based on misinformation. Much of the recent hostility was based on superficial perceived slights, what Outlaw and Butler called "face fighting": occurrences as trivial as the way one kid looked at another on the street. Outlaw told the kids he knew many people who had been killed because of face fighting. "Imagine that," Outlaw scoffed, "being killed because you can't let go of the way some motherfucker looked at you. You the ultimate sucka if you die like that." By the end, Outlaw and Butler got the boys to shake hands and agree not to go after each other. Unlike earlier truces, however, this time Outlaw didn't concern himself with written documents for the kids to sign. "For real men," Outlaw and Butler said, "their word is their bond. For us, you don't gotta sign shit. But if you say it, you better back it up." Afterward, Outlaw and Butler brought the two crews, separately, to the Apollo Theater in Harlem and then, together this time, to a movie. The truce held up for months and then years. Ten years later, the truce still holds. "It made a big difference," said Lieutenant Thaddeus Reddish, a police officer in Newhallville, about the intervention. "It really brought the level of violence down."

In 2011, not long after the truce, Doug, Barbara Tinney, and Outlaw were joined by a fourth essential member, Pepe Vega. Pepe was forty, although he didn't look it. Fit and sporting the latest hip-hop fashions, he resembled a recently retired boxer. Pepe had multiple dimensions: he'd written screenplays and produced low-budget movies about gang life, and had a side business running a sneaker store. He was nearing completion of a bachelor's degree in business. When he was younger, he was a mainstay in a gang in New Haven and a prolific drug dealer, although he never used them, hating what they did to his body. He'd be shot twice before he finally left that life behind. Pepe's initial exposure to the criminal justice system came at a tragically young age. His father, incarcerated for assault, hanged himself in the New Haven jail when Pepe was six months old. To this day, neither Pepe nor his mother knows quite what happened in the cell that night. When Pepe joined the outreach workers, the team embraced his focused and disciplined energy. The fact that he spoke Spanish too was critical—he served as an entrée to a whole new set of families.

From the first day on the job, Outlaw fell in love with the idea of being a street outreach worker. It reengaged all of his talents: sizing up situations, strategy and organization, leading young people, and problem solving in the midst of crisis and threats of violence. He loved being back in the middle of everything and being able to use all the same skills he had used as a Jungle Boy, but this time for prosocial purposes rather than antisocial ones. Outlaw couldn't believe his good fortune: he was a gang-leader again.

As part of the initial city contract for the outreach team, Yale's Robert Wood Johnson scholars were hired to evaluate the early impact of the program. Yale's conclusion was that the team was "successful beyond expectations." Youth engaged by the team were significantly more likely to have attended school (79 percent versus 35 percent) and to have been employed (53 percent versus 14 percent), and were significantly less likely to have been incarcerated (19 percent versus 56 percent). Most critically, however, youth engaged in the program were significantly less likely to be victims of shootings (7 percent versus 26 percent). A subsequent study showed that in the first four years of the team's existence, nonfatal gun violence between youth was reduced by a third. In 2012, for example, the team intervened in 160 potentially violent disputes involving 645 youths. When the City of New Haven received a major federal anti-crime grant, the city council agreed to direct the funds toward the outreach team in order to expand their roster and broaden their services.

But even in the midst of these successes, controversy continued to surround the team. Many in the police force remained skeptical, and some residents believed that team members were tipping off gang kids about the police's interest in them. There were rumors around town that Outlaw's newly respectable image was the ultimate cover for what he was actually doing—running a major cocaine and heroin ring. Beyond the conspiracy theories, there were real problems on the team—specifically, a very high turnover of staff, a product, no doubt, of the job's unique combination of high stress, danger, and low pay. A number of team members were fired for drug relapses and rearrests, as well as for threatening behavior to others outside of the job, although no violence ever actually occurred. And certainly, even at baseline, the team did not often adhere to conventional models of professional decorum. Hysterical laughter, heated arguments, and simply loud talk—often coming from Outlaw—dominated their weekly staff meetings. Staff members at New Haven Family Alliance would shut the conference room door on the team, and talk under their breath about how volatile they appeared, teetering on the edge of control.

Although the outreach team felt affirmed by the outcomes of the evaluation studies, they knew the reports represented only part of what they had achieved. To a person, the workers understood the studies didn't capture what was arguably more important than what had happened: what did *not* happen as a result of their work. Outlaw knew that to be on the team meant knowing how to operate in and master the world of counterfactuals.

In 2012, the best friend of one of Outlaw's clients was killed at age sixteen, caught in the crossfire between two gang members. Outlaw's client, John, contacted Outlaw on the night of the murder. Everyone on the street knew who did it, and John was going to kill him. Outlaw understood John's thirst for revenge. The murdered boy had been a joyful, fun kid whom everyone loved; his family was destroyed; and the boy, like Jajuana Cole and Justus Suggs, had simply been in the wrong place at the wrong time. But Outlaw told John that night, "I know you want to represent your boy, but you gotta find a positive way to do it. Not a negative one. Do something good with your life to let your friend's legacy live on." Even in his grief, John heard Outlaw out that night and chose not to retaliate. John wound up finishing high school and went to college. He is now a sheriff in the courthouse where his friend's murderer was convicted.

On another occasion, Outlaw got a call from a distraught mother in Newhallville. The mom was crying and needed Outlaw's help. She was despairing over her teenage son, who she knew was newly involved with a gang. The mom said she had something she needed to give to Outlaw, and the police would need to be involved. Outlaw called Doug, who had by far the best relations with the police department of any of the outreach workers, having worked there for a decade. Doug, Outlaw, and the mom arranged to meet on Dixwell Avenue at two in the afternoon, in broad daylight. The mom handed Doug a black gym bag. Outlaw and Doug knew what the bag contained, as the mom had told them previously.

Doug went to the nearby police substation, asked for the district supervisor, went into an office, opened up the bag, took out a gun, and turned it over to the officer. For years afterward, whenever Doug and Outlaw ran into the mom around town, she effusively thanked them. She admitted she had taken the gun from her son without his knowledge. At first the boy was angry, saying she had left him vulnerable to being killed, but later he was grateful, especially when he found that it had been Outlaw who had been behind the move. The fact that Outlaw and Doug had backed up the arrangement meant that he wasn't being weak, or "a pussy," in the code of the streets. The boy left the gang world behind.

As well as Outlaw was doing professionally, other parts of his life were under constant strain. He was chronically short of money. Joshua, the boy that he had fathered shortly after being released from prison, was now two years old, and he was delayed in his speech and often unresponsive to social cues. Outlaw and the boy's mother took Joshua to a child guidance center where he was diagnosed with severe autism. Outlaw enrolled Joshua in a five-day-a-week intensive program for children with early-stage autism, which Outlaw attended one day every two to three weeks to help out. Pearl, now retired, was beginning to have mobility problems and difficulty driving, and was in and out of nursing homes. Outlaw was also engaged in the long and difficult, perhaps impossible, process of attempting to mend relationships with his five now adult children. As much as he might try to compensate now, he could never undo the fact that he had been unavailable for his children when they were growing up. Outlaw's children were often angry and distant, and now sometimes resentful. Outlaw may have been celebrated for helping kids he barely knew, but he hadn't even raised his own children.

Outlaw also arguably suffered from symptoms of post-traumatic stress disorder, or what he called "postincarceration syndrome." Getting out of a parked car, he checked all angles for potential lines of attack and gunfire, second nature from years of prison and as a gangleader. In every professional meeting, Outlaw sat close to the exit in case he needed to make a quick escape. His sleeping problems never improved. He was at times atypically irritable, was easily startled, and walked around with red, irritated eyes from washing his hair in the shower with his eyes open.

IN 2011, OUTLAW got into a major family dispute with his uncle. The argument occurred in Pearl's apartment after a wake for Pearl's sister, who had passed away. For the previous twenty-three years, the uncle and his family had been living in a house that Outlaw had bought when he was a Jungle Boy, and then given to the uncle "on loan" when Outlaw was incarcerated. Now that he was out, Outlaw had asked for the house back; he was tired of living in apartments.

"You ain't been here for twenty years, and you want it back?" the uncle said.

"Yes, I do," Outlaw said. "It's mine."

"No, it ain't," was the response.

"Fuck you," Outlaw said.

The uncle was incensed. "You know what else, Juneboy, who thinks he knows everything? You know what else?"

"You tell me," Outlaw said.

"All right, then. I'll tell you something. I'll tell you something big. Your daddy ain't your daddy. Your mom slept with someone, got pregnant with you, and that's why the guy you thought was your daddy your whole life left."

"Then who is my daddy?" Outlaw said.

"Robert Earl."

"Who?"

"Robert Earl."

It took a moment for Outlaw to place the name. Robert Earl. He racked his brain, and then it came to him: Robert Earl had lived in a house five doors down from where Outlaw first started dealing on Lilac Street. Outlaw recalled seeing him from time to time on the corner of Lilac and Starr. He was an older guy, a big, gruff man, known in the neighborhood as a heavy drinker, but a guy who could hold his liquor and a workingman. Outlaw knew him enough to say hello, but that was it. He also knew that Robert Earl had died years ago, when Outlaw was away.

Outlaw stumbled out of the wake in a daze, feeling equal parts bewilderment and anger. Pearl had said nothing during the exchange, confirming, most likely, the story's credibility. Outlaw knew not to ask her any questions about the matter: he knew she wouldn't answer them. He went where he always did in crisis, the slab of concrete overlooking the Quinnipiac River—his "bunking out" spot. Looking out at the gray water, he ransacked his brain for any other memories of Robert Earl. One finally came to him: a curious encounter he'd had with him when they'd both been held for a night in the New Haven jail. Outlaw was in for trespassing or marijuana possession or some other easily bailed out charge, and Earl for public nuisance or something like that. Earl's charge wasn't serious: everyone knew he wasn't truly a criminal. Outlaw recalled that during that night in jail, Robert Earl had given Outlaw a photograph of Robert with his brothers. Outlaw accepted the picture, but it seemed like an odd gesture at the time. Now, recalling the incident after twenty-five years, he wondered: Did this mean that Robert Earl knew that Outlaw was his son? Was the gift of the photograph Robert Earl's way of sharing with Outlaw his real father and his uncles?

Down by the river, Outlaw skipped stones over the water. The rhythmic simplicity of the motion soothed him. Outlaw tried to fathom that the man who he'd always believed was his father was very likely not his father. What did that mean? Was he not really an "Outlaw" anymore? Should he have a different last name? How would his life have turned out if he'd known as a kid that Robert Earl was his father? Maybe he would have gotten some actual attention from a father and his entire life would have been different. But after an hour of skimming stones across the expanse, Outlaw slowly came to the conclusion that there was nothing he could do about the situation. Robert Earl was dead, and Pearl would never talk.

As Outlaw remastered New Haven, he noticed that many things had gotten worse during the time he'd been away. Poverty and violence now pervaded entire neighborhoods. In the Jungle Boys era the pathology seemed more isolated to specific blocks and specific segments of the population. Indeed, a broad survey in 2010 of middle school youth in *all* city neighborhoods found that a third had seen someone shot or stabbed, 1 in 10 reported having carried a gun or a weapon, almost 2 in 10 had been involved in a gang fight, and a quarter had hurt someone badly in a physical fight. A study by Yale's School of Public Health of 1,000 residents in 6 low-income neighborhoods in New Haven found that 3 in 4 had heard gunshots, and 1 in 5 had a family member or close friend who had been killed violently.

Having spent time in the homes of many of their clients, Outlaw and other members of the outreach team estimated that half of the kids went hungry on a weekly basis. Food insecurity was actually one of the most vexing obstacles that the team faced, maybe even the biggest. It was nearly impossible to engage a kid with job training, for instance, with the promise of employment three months into the future, when they were starving today. For that matter, it felt almost cruel to tell a kid not to deal drugs when they needed to eat. Yet, even with food lacking in the home, Outlaw found that many of the apartments reeked of marijuana.

Outlaw also found a gang world that had been utterly transformed. With the arrival of social media, many of the local gangs were no longer site-based, as the Jungle Boys had been. Nor were the gangs the tight, highly organized, small corporate entities they were in the 1980s but rather small, loosely affiliated, and fluid groups. The kids in New Haven were connected via Facebook and Instagram to gang members across the country, from whom they got hand signs, rules, even handbooks. But what this really meant in practice, Outlaw found, was that a sixteen-year-old "wannabe" gangster in Newhallville would download slogans and pretend for a week or a month to be a Crip or a Blood. Outlaw thought it ludicrous. His gang had been homegrown, not something imported from Los Angeles. That so many of the gangs were, in effect, virtual also made it difficult for the outreach team to contact them and track their activities. A given crew was comprised of members from all over the city, or even the suburbs, their movements and communications managed through cell phones and social media. In 2016, when a New Haven dealer was sent back to prison, he sold his cell phone to another dealer for $30,000. The contacts on his phone amounted to an instant book of business.

Social media also added to the contagion of violence. Gang members used Facebook and other platforms daily to feud and make threats, post about their newest weapons and express their affiliations, and, of course, sell drugs. After a killing or shooting, word (and pictures) spread throughout the town in minutes. The public nature of these postings often led to increased social pressure for retaliation. Kids often posted pictures related to their crimes, allowing the police to catch them with ludicrous ease.

Even the drugs had changed. The appeal of crack and cocaine had dissipated during the time Outlaw was in prison, replaced by opioids, prescription drugs sold on the street, crystal meth, and designer drugs that didn't even exist in the 1980s. The profits were a fraction of what they used to be, the street value of cocaine and heroin now about a quarter or even a tenth of what it had been. Many of the changes in gang life dismayed Outlaw: at least in his day, there'd been something worth fighting for, lots of money and the loyalty, if not love, of his fellow soldiers. He encountered hopelessness and nihilism on the streets that he didn't recall among at least the more established Jungle Boys. Alfalfa, Pong, Rodney, Outlaw: they all would have died for one another, and though they would never have admitted it, they loved one another.

Outlaw began to perceive the gangs as a kind of ever-changing and multiheaded monster. He found that he could get the leaders in line but then the younger "hotheads" underneath would spring up. Months of work to negotiate a truce or détente could be subverted by the aggression of junior gang members hungry to make their mark. To Outlaw and Doug Bethea and the rest of the team, the violence always came down to fifty or seventy-five hard-core guys. Outlaw estimated that there were two hundred gang-involved kids in town but the majority of them were wannabes, followers, or as he called them, "punks behind guns," who were influenced, cajoled, and intimidated by the hard-core guys at the center. Often, however, Outlaw found that the individuals most at risk were those who were not fully committed to the gang life—those who were ambivalent, half in, half out of the underworld. Being in a gang at least offered protection.

In order to adapt to the new terrain, Outlaw had to master contemporary tools. He quickly became immersed in the digital world to which his clients were constantly connected. Outlaw's work became centered on his two iPhones, one for work, one for business, which he used—and dropped—so frequently that they both had broken faceplates. He conducted major negotiations via quick conversations with kids on the phone, which, given their brevity and use of insider slang, were often unintelligible to outsiders. Outlaw constantly posted strategic messages on his various Facebook accounts to his contacts, which now numbered in the many thousands. At the beginning of the summer—that season typically being the most violent time of the year—Outlaw always went on Facebook Live.

"Yo, man, I haven't gone live for a while, but I wanted to address the town. This is to the kids out there. We heading into the summer and people need to stay cool. I don't want nobody to retaliate against nobody. If you feel something been done wrong to you, you come to me or the guys on the outreach team, you know what I'm saying? . . . If there are any beefs out there, you gotta fuck the beef. You gotta ice the beef."

Within minutes, his videos were seen by hundreds of people, galvanized by his message:

"You tell them, JB!"

"Respect."

Emojis of brown hands in a prayer position.

Outlaw ended his ten-minute speech with, "I make it look easy, but it's hard. It ain't easy doing this shit. The violence is hard. That why we got to stop it."

But there was one thing that had not much changed upon Outlaw's return: the homicide rate. Twenty-two murders were perpetrated in New Haven in 2008; 12 in 2009; 23 in 2010; and an astonishing 34 in 2011, numbers that had not been seen since the 1980s. In 2010, not one of the 24 murder victims in New Haven was white: 23 were black, and one was Hispanic. Twenty-three of the 24 people killed were male. Between 2003 and 2015, among the 1,225 victims of gunshot wounds treated at Yale New Haven Hospital, 84 percent were either black or Hispanic, and 93 percent were male. The statistics mirrored national numbers: black men comprise 6 percent of the United States population but half of all homicide victims. Five thousand black men perish in gunfire annually.

One murder in 2009 was atypical in that its victim was neither male nor black. The day before she was to be married, the body of Annie Le, a twenty-four-year-old Yale doctoral student, was found behind the panels of a wall in a university laboratory building. The story made international news, and a vigil with two thousand participants was held on the Yale campus. A hundred investigators were assigned to the case by the Yale and New Haven police departments, the state police, and the FBI, and Yale offered a cash reward. Le's forensic autopsy was performed immediately at the state police laboratory, jumping to the front of the months-long backlog of cases. Within forty-eight hours, DNA matches from Le's remains were linked to a technician in the lab where Le was found. He was arrested nine days later for the murder.

In the ten years since Annie Le's killing, more than fifty people have been murdered in New Haven. Virtually all of the victims—and the perpetrators—were black men. Many of these murders are unsolved. All of these murders put together have received a fraction of the media coverage and attention of the Annie Le slaying.

To Outlaw, and many others, the story remains unchanged from when he was a gangster: black men being killed, and the community largely accepting it. But Outlaw doesn't feel that the New Haven police department, these days, anyway, is fundamentally racist. In the days after a rash of killings of unarmed black men by white police across the country, he posted on Facebook: "New Haven doesn't have a police brutality problem. It has a black on black violence problem." Outlaw feels the reason that so much attention was paid to the Annie Le murder, and not to those of his many clients, is simple. Annie Le was killed where the money is: Yale.

At the heart of Outlaw's work is this notion of being an interrupter: of stopping things that might have happened from actually happening. Jumping in at exactly the right time makes all the difference.

In the summer of 2016, Outlaw set up a meeting on Howard Avenue, a few blocks from Walter Brooks House, to meet with Patrick Hale. Outlaw arrived early and bought an Italian ice from a street vendor. The ices used to be sold by Italians back in his day; now it was a Hispanic woman, doing a good trade.

Patrick Hale sauntered up to Outlaw at 2 P.M., exactly at the appointed time. He was light-skinned, lean and strong, wearing blue jeans and a hoodie. Outlaw offered to buy him an Italian ice. Patrick declined. Looking into Patrick's face, at his alert eyes, Outlaw saw a bit of himself in the young man. He was bright, aggressive, muscular, one of the up-and-coming dealers in town. Outlaw respected him for having a real crew.

They greeted each other: "How you doin'? What up?" The reason for the meeting was straightforward: Patrick had just been arrested on a violation of probation charge and in two weeks, at age twenty-two, he was going to be sentenced to prison for the first time. Outlaw was going to give him some practical advice on how to survive being locked up. Outlaw told Patrick that he had been in twenty different prisons in the course of his career. He was just starting to tell Patrick how to get by behind bars— keep your head low, don't talk unless you're spoken to, don't join a gang, don't mouth off to the corrections officers because at the end of the day you are not going to win—when a beat-up red Ford sedan slowed down and pulled over. The car badly needed a muffler. The driver rolled down the passenger-side window, and called out to Patrick and Outlaw. Outlaw recognized the driver. Andre. He was a mouthy kid, known around town for doing stickup jobs: muggings, and robberies of bodegas and small-time dealers. He was a lone and unpredictable operator, not capable or likable enough to be part of a gang.

"What up," said Andre, a malevolent edge in his voice.

Patrick said hello. There was an awkward pause. Patrick had an unlit cigarette in his hand. For lack of anything better to say or do, Patrick asked Andre for a light. Andre said, "Sure, I'll get you a light," the edge in his voice still there. Outlaw watched as Andre pulled out a gun instead of a lighter and flashed it directly at Patrick. After waving the revolver recklessly in Patrick's direction for a couple of seconds, Andre floored the car and tore away down the street. The tires burned. Patrick looked at Outlaw, incredulous.

"What the fuck? I can't believe he just did that."

Outlaw tried to calm Patrick down.

"He's just a punk, a nobody," Outlaw said. "Don't worry about it."

But Patrick wouldn't back down. "The motherfucker disrespected me. I'm going to blow his fucking lid off." Outlaw tried again to soothe or distract Patrick, but he couldn't be dissuaded. Outlaw gave up on the prison conversation and Patrick finally said that he was too pissed off and had to go. "Don't do anything stupid," Outlaw told him.

Later that afternoon, Outlaw called Andre's friends and associates. He told them to get Andre the following message, immediately: "You got to seriously stay low. Yo, you really shouldn't go out tonight." Later, Outlaw repeatedly tried to reach Patrick but got no answer. Outlaw called Patrick's crew. He asked them to try to settle Patrick down, take him out to a bar or something, anything to take his mind off Andre.

Later that night Outlaw went to his condo but couldn't sleep. He couldn't stop thinking about Patrick and Andre: *Could I have done something different? Was someone going to get hurt, even die, tonight?*

The next morning, Outlaw finally got through to both Patrick's crew and Andre's associates. Through their intermediaries, he arranged a meeting for later that night. Outlaw picked the time—seven-thirty—and the place—a children's playground, Roberto Clemente Park in the Hill. He chose the site with care. It was a fairly quiet place, especially at that time of the evening, so that passersby would likely not notice that they were meeting. But neither was it totally isolated. Enough people would be around, walking past and driving through, that if anyone tried something stupid it would attract attention. Outlaw picked one of the best outreach workers, Roberto, to come along. Roberto didn't know Patrick or Andre or their circles, and Outlaw thought this might prove an advantage: they might behave better around a stranger.

Patrick and Andre appeared on time. They both came alone. Roberto patted them down. No guns. Neither Patrick nor Andre knew that the other was going to be there. Outlaw hadn't told them. They were surprised and scared, especially Andre. Outlaw saw the fear in his face.

Roberto stood by as Outlaw did most of the talking. "This is real serious," he began. "Everybody knows why we're here. And the main thing is: we don't want nobody to go to jail."

At first Patrick and Andre didn't make eye contact. They looked at their feet but nodded when Outlaw spoke. Outlaw asked them, "What's the problem? What's the beef?"

Patrick started off by saying, "Why'd you disrespect me? Why did you choose me? I don't want no beef with nobody."

Andre began to apologize, and once he started, he couldn't stop. Patrick heard that, and appeared to appreciate it. As they talked, it turned out there was a history between the two of them, one that neither of them had previously fully comprehended. It came out that Andre had stolen a gold chain from one of Patrick's crew members and pawned it. When Outlaw heard that, he told Andre to get the chain back to Patrick as a form of apology. Andre, still worried for his life, agreed. Indeed, Andre went to the pawnshop the next day. Remarkably the gold chain was still there; he bought it back from the pawnbroker and returned it to Patrick.

A few weeks later Patrick went to prison without incident, and without causing any further trouble. Outlaw thinks that Andre still doesn't fully realize how close he had been to getting killed.

As this conversation took place in the Hill, three-quarters of a mile away at Yale, the sociologist Andrew Papachristos was in the midst of a multiyear study on the factors that underlie gunshot victimization in the United States. Papachristos examined nonfatal gunshot injuries in Chicago over eight years and found the incidents were heavily linked to the social networks in which suspects and victims traveled: the more an individual was embedded in a social network that included gunshot victims, the higher the risk of their own victimization. Seventy percent of the killings in Chicago occurred within social networks that comprised just 6 percent of the population. Furthermore, victimization from gun violence occurred through a process of "social contagion," which Papachristos defined as "the process by which people in the same social network profoundly affect each other's feelings, ideas, and behaviors." As a measure of social ties and networks, Papachristos studied tens of thousands of co-offenders, meaning individuals who were arrested together for the same offense. He found that two-thirds of gunshot victimizations were explained by social contagion. If one co-offender was shot, the other co-offender was shot on average within 125 days afterward. Papachristos uncovered extraordinarily intertwined social networks among gunshot victims, which he called cascades. As he put it, "I get shot, then my associate gets shot, and then my associate's associates get shot." He discovered 4,107 such cascades in Chicago, including one that started with a single incident that was linked through the victims' social networks to 469 separate violent events.

Papachristos's work may lead to new ways of thinking about violence prevention. "Tracing violent episodes through social networks could provide valuable information not only for law enforcement," he says, "but for public health and medical professionals to create effective interventions with people and communities at highest risk." In other words, if groups like street outreach teams infiltrate social networks, and successfully disrupt or interrupt the process of contagion, they can save lives, perhaps many lives. Stop one murder, and you could stop six more. Stop one shooting, and you could stop sixty, over the course of years.

Outlaw is unfamiliar with the work of Andrew Papachristos, but he knows as well as anyone in New Haven how social networks operate, and how cascades of violent events unfold. After all, as a Jungle Boy, he was responsible for creating cascades of violence. By stopping Patrick from shooting or killing Andre, he could have saved multiple lives.

OUTLAW'S BOSS SINCE 2015, and the current leader of the outreach team, is Leonard Jahad, who was born Leonard Jackson. Jahad is the only member of the team who is conversant in Andrew Papachristos's work and can translate it for the team. In fact, Jahad, in his former position as chief of probation in New Haven, served on a number of committees with Papachristos. Jahad tells the outreach workers to think of themselves as public health workers who stop an epidemic, but in this case the epidemic is violence.

Jahad, as everyone in New Haven calls him, is at age fifty-two a year older than Outlaw. At six foot three and 250 pounds, he has the height and heft of a former football star, which he is. His head is shaved, and he frequently breaks into a gap-toothed smile. Jahad grew up on the outskirts of New Haven in a middle-class Baptist family. His parents, like Outlaw's, came from the South—in Jahad's case, to escape the cotton fields of Georgia. He grew up in a disciplined household. Even though the family could afford heating oil, Jahad was told by his father to chop wood daily in the winter for extra provisions. In high school, Jahad was a fine student and athlete. Upon graduation he went to an all-black college in Atlanta, where he played football until sophomore year, when he was kicked off the team, and out of school, for a fistfight with another player. It was an unmoored time in his life. He returned to New Haven and got involved with selling drugs, working as a "bag man"—dropping off drugs and picking up money—for Jamaican gangsters, an offshoot of the posses with whom Outlaw warred. But Jahad never felt comfortable in the role.

A younger white policeman pulled him over one day. "What the fuck's wrong with you, Leonard?" Jahad had never met the officer before. "How do you know my name's Leonard?" Jahad asked. "I know because the Jamaicans already sold you out. They told me all about you." Jahad was speechless. "Let me tell you something," the officer said. "You're not made for this. You got 'college' written all over you."

"Let me ask you," the officer continued, getting out of his car, "what are your goals in life?" Leonard felt uncomfortable with the way the conversation was going, but the policeman persisted. "Do you want money?" "Sure," Jahad said, smiling uncomfortably. "And you want respect, right?" Jahad nodded. "Do you like guns? Do you like fast cars?" Jahad nodded again. "Well, you know what, you should be a policeman," the officer said. "You can get all those things, just the same as being a criminal. But you sleep better at night."

The next day, Jahad applied to be a corrections officer in the New Haven jail, where he worked for the next ten years, until he transferred to the state probation department. He much preferred working with clients in the community and was skilled in his rapport with them. His clients would come into the office and he would promptly take them out to the pizza place across the street. His colleagues teased him about this, but everyone acknowledged that Jahad knew more about criminal activity than any of the other officers. "Where do you think I'm getting all that street intel from?" Jahad would retort. "It's the pizza!" Eventually he was promoted to chief probation officer. In the meantime, he had converted to Islam, became a foster parent to a number of children, coached youth football, and was a leader in the New Haven Masons organization. Many kids around New Haven called him, simply, "Dad."

During Jahad's tenure as chief of probation, a recently retired policeman came into the office seeking part-time work as a probation officer. "I always wondered what happened to you," the officer said. Jahad was confused, as the man didn't look familiar. "I'm John Bashta. I was the policeman who told you to get off the streets all those years ago. And now I'm asking you for a job!" Needless to say, Bashta was hired. (John Bashta was the officer who first heard the gunshots that killed Sterling Williams and found Williams dead on Church Street South.)

Jahad, like virtually every African-American person of his generation in New Haven, knew Outlaw when they were growing up. To Jahad, Outlaw was a "keep it moving" kind of guy: "You say hi and then move along fast. He was trouble." It had been twenty-five years since Jahad had seen Outlaw. On the street Jahad greeted him warmly but maintained a wary distance. Jahad was doubtful that Outlaw could actually be reformed, until he saw him give a speech to some high school students. Outlaw began the speech by apologizing to his victims: to the people he had shot, to his family, to the people he'd sold drugs to. Jahad had worked with a lot of criminals, and reformed criminals, but he'd never seen anyone speak in such a powerful way before.

Outlaw is now Jahad's right-hand man. When Jahad is on vacation, Outlaw runs the meetings. "Juneboy," Jahad says, with his gap-toothed smile, "he's my man."

But sometimes, despite all the good work by Bethea and Jahad and Outlaw, the violence isn't interrupted but proceeds apace, and no amount of counseling and goodwill and mentoring stops the bullets from leaving the chambers of a gun.

Billy Lyons was one of Outlaw's favorite clients. A good-looking kid with a smile that could light up a room, Billy was, as Outlaw put it, "no slouch." Like Outlaw, Billy grew up in a generally middle-class family in Newhallville. He was bright and hungry, and Outlaw saw something of his young self in Billy. Billy's ambition and nerve led him down the wrong avenues: he was allegedly an active member of a Bloods-affiliated gang. His mother tried to keep him from that life, including sending him down to Pennsylvania for a few months. Billy returned and vowed to give up the gang life for good. "He was so happy the last time I saw him," Outlaw said. "He had just gotten off of work at Wendy's. He was so happy about how much progress he had made." Four days later Billy was found slumped over in the front seat of his car, shot in the back of the head. A pedestrian noticed him draped over the steering wheel, blood everywhere. Billy was seventeen. It came out after his death that he had been a marked man. Outlaw arrived at the murder scene not long after the police found the body. The area around the car was cordoned off, and officers and emergency responders were everywhere. Outlaw, broken up himself, hugged Billy's mother, who was apoplectic with grief, and attempted to console other family members. Billy's mother had lost not only her son but also a cousin and a nephew to shootings in New Haven. Outlaw functioned at the crime scene, but afterward, in his condo, and later at the funeral, he grieved yet another boy lost to the streets, another kid with talent. Outlaw knew he could easily have met Billy's fate five times over.

A number of years later one of Outlaw's clients was convicted of Billy's homicide.

Outlaw had counseled the man convicted of killing Billy multiple times and thought he had successfully engaged him, and made an impact on him. "He was doing well, working at a decent job. But I remember once when he challenged me. I was talking about how I had changed. He said back to me, with real emotion, 'Change may be easy for you. But I think it's hard for me.'

"He looked real disturbed when he said that," Outlaw said.

CHAPTER 9

The Maintenance of Hope

Outlaw and his fiancée, Germaine, 2019.

WHEN SHE FIRST MET HIM, GERMAINE LYDE WAS SCARED OF WILLIAM Outlaw. She wasn't so much concerned about his Jungle Boys past, although she had heard plenty of stories, but rather about his ability to stay faithful. When Germaine and Outlaw first started going out, he had been open about his former promiscuity. "I don't have a great track record with women," he admitted. "When you've slept with three sets of mothers and their daughters in New Haven, you know that ain't good."

Germaine, a professional, hardworking Christian woman with a college degree, had worked for twenty years as a counselor in social service agencies. She lived in a comfortable townhouse and had a grown daughter in graduate school to become a social worker. Her quiet, predictable life revolved around work, her daughter, and church. The last thing she needed in her life was William Outlaw.

She first met Outlaw when he worked at Dunkin' Donuts, but they got to know each other well sometime later. A year into his tenure as an outreach worker, Outlaw felt that he had mastered the job sufficiently that he could do more in the community. He also needed more money to help his kids and now grandkids. In 2011, Outlaw took a second full-time job as a case manager at Goodwill Southern New England, where Germaine was working as a counselor. He worked nine to four at Goodwill, and five to midnight on the outreach team. Goodwill Southern New England is the regional branch of the national organization, which specializes in employment training for people with disabilities. The New Haven office had recently developed a new service line offering training and support for recently released offenders. (After all, this was a growth business: as the nation incarcerated more people, more individuals were released from prisons and jails.) At Goodwill, Germaine mainly kept to her cubicle, meeting individually with clients to prepare résumés, conduct mock employment interviews, and obtain interview-appropriate clothing.

Just as when he joined the outreach team, Outlaw immediately immersed himself in the work, pounding the pavement and knocking on the doors of trucking companies, warehouses, fast-food places, and mom-and-pop restaurants: any company willing to hire people just out of prison. Soon enough, Outlaw greatly enhanced the local network of such employers and created a list of job openings for clients, which he distributed weekly and posted on Facebook. With a yellow marker, he highlighted the positions and their pay rates, and wrote: "No excuses! Get busy! Hustlers got to hustle."

Around the office, Germaine was as quiet and methodical as Outlaw was extroverted and entrepreneurial. After a few months of flirtatious watercooler talk, Germaine—despite her better judgment, she sometimes thought—began to date Outlaw. Concerned about getting fired, they kept their office romance discreet, but on Friday and Saturday nights they'd go out to dinner and the movies. Outlaw would bring flowers and chocolates. When he learned that Germaine liked lilies, he drove to florists all over town in order to find the freshest ones. He would pamper her on her birthday and Valentine's Day, driving her to Harlem to eat at a steakhouse and dance at a club. Germaine began to think of Outlaw as a big teddy bear. After six months of respectful courting, Outlaw moved into Germaine's townhouse in Fair Haven, just a few blocks from where he'd grown up at Q View. Outlaw paid more than his share of the rent, and did a lot of the cooking and shopping and cleaning. He embraced a domestic life that he'd never known before. He also visited his son Joshua, who was now four years old and living with his mother, almost daily. Joshua was making good progress at his school for children with special needs.

At Goodwill, where Outlaw is called "Will" or "Big Will"—and not Juneboy—by his mainly white and middle-class colleagues, who know little of his former life, Outlaw was quickly promoted to a supervisory role. He began spending parts of two or three days a week in prisons meeting soon-to-be-released offenders and preparing them for the psychological, financial, emotional, and familial demands of returning to society. Many of Outlaw's visits bring him back to Osborne, formerly Somers, and a surge of anxiety and nausea greets him every time he passes through the gates. Often he picks up prisoners on their day of release, either at the gates of the facilities in northern Connecticut or at the New Haven jail, where they are dropped off by parole officers at 6:30 A.M. on what is called "court run." He takes his clients out for a cup of coffee and a drive around town as their first experience as free men and women, and sometimes brings them to the office to help them find a place to sleep that night.

As part of his duties at Goodwill, he runs a weekly Men Helping Men group for just-released prisoners in one of Project M.O.R.E.'s halfway houses. The room is packed with twenty attendees every Thursday evening at five. The men, many of whom are still wearing prison garb because they can't afford street clothes, receive a much-needed pasta and salad meal and a bus token for their attendance. The Department of Correction parole officers encourage all their New Haven–area clients to attend, and in some cases mandate attendance, sensing that some of the highest-risk individuals need someone of Outlaw's gravitas as a mentor. Ed Kendall, the former police detective who butted heads repeatedly with Outlaw in the 1980s and was instrumental in dismantling the Jungle Boys in 1992, is now in charge of the mental health unit of the New Haven parole office. He insists that all his clients attend Outlaw's group. "I can't say enough about it," Kendall says. "Young guys, old guys, and black guys, white guys: they all tell me it's the highlight of their week and it keeps them on the right path."

Outlaw announces at the start of each week's session that the group has one goal and one goal only: "to reduce recidivism and keep you guys out of the fucking penitentiary." As the group members introduce themselves, Outlaw warns them not to use their street names, nor to talk about what neighborhood they are from, their criminal pasts, or how much time they spent in prison. "Don't tell no war stories," he says. "The present is all that matters. Remember, it's never too late to change, to get an education, and be a better father. Believe me, I know." Outlaw bases much of the group's focus on his own experience with the Lewisburg therapist Richard Whitmire. Often he has the group do exercises modified from Whitmire's workbook.

Bob Doucette, a white man in his early sixties, with lurid turquoise and orange tattoos that travel down his arms and up his neck, is the mainstay of the group. He arrives an hour early to lay out the food, soda, napkins, and cutlery. He has attended every week for five years. After decades in and out of prison for drug charges and robberies, Doucette was initially mandated by his parole officer to attend the group for six months, but he has attended voluntarily since then. For Doucette, Outlaw has a way of "keeping it real. And he always believed in me." For the first time since he was fifteen years old, Doucette is drug-free. Crack cocaine was his drug of choice and he's been clean for the last five years, which is the exact time he's attended the group. Bob lives in a rooming house and now has two jobs: as a dishwasher at a café and stocking the shelves at a dollar store. In the spring of 2017, he texted Outlaw, "I lost my mom

2 days ago. Didn't think it would be this hard to deal with. She told me before she passed away that she was proud of me and she made me promise her that I would continue to do the right things in life. Keep me in your prayers pls." Outlaw replied, "Stay strong brother." The funeral was at the veterans' cemetery in a town twenty miles from New Haven. The only nonfamily member at the funeral was William Outlaw. "Thanks for showing up Friday. My family wanted me to relay that to you," Doucette wrote Outlaw afterward. "Everyone thought you were my parole officer."

Outlaw finds that his two full-time jobs at Goodwill and with the street outreach team mutually reinforce each other, often to powerful effect. On prison visits, Outlaw often runs into young men whom he had attempted to counsel on the streets just a few months before. "Yo, JB, you were right!" they say, the shame and embarrassment shining in their eyes. "You told me that if I kept up the street life, that this was going to happen. I should have listened to you." Outlaw tries not to rub it in, but he doesn't disagree. He *had* tried to warn them. But then again, various people, from Coach Saulsbury to Pearl, had tried to warn Outlaw when he was a teenager, and he had paid no attention.

At Goodwill, Outlaw is especially close with Mary Loftus, a sixtyish Irish-American woman with a kind manner and droll sense of humor. She is a senior case manager and has a lifetime of good works behind her, as a counselor and attendant in shelters for the homeless and domestic violence victims. Outlaw and Mary share an office, and their days are filled with warm banter and mutual productivity. On her birthday, Outlaw posted on Facebook: "Happy Birthday To My Coworker, Friend, Chef, Boss, and The Best Casemanager In Conn. No disrespect to all the other casemanagers, It's Just My Love I Have For Mary Loftus =100%OFFICIAL." A few years ago, Mary's daughter died of cancer in her thirties. "Will helped me so much during that time," Mary said. "When I needed to vent, he let me vent. He covered my job for months. When a guy needed work clothes, he went shopping with them. When I couldn't do a court run at 6 A.M., he went to pick up my client. Knowing that he had my back made all the difference." Outlaw has a teddy bear on his desk, a gift from Mary's daughter before she got sick. Outlaw has loving relationships with others in his office. He lends money he doesn't have to fired employees and when an officemate, a woman in her thirties, had an unexpected stroke, Outlaw visited her in the hospital and sent her a card: "God loves you. Try not to worry so much."

In 2015, at the annual meeting of the two hundred employees of his branch of Goodwill in the ballroom of a suburban hotel, Outlaw, to his utter shock, won the Employee of the Year award. He stood at the podium next to the CEO as a standing ovation went on for two minutes. In those moments, Outlaw saw images from his former life pass by him: from Pearl's red Buick to fist fighting at Q View, from duffel bags of cash to firing guns, from *Once Upon a Time in America* to the toilets flooding in Building 63, from Termite to Frank James to Warren Kimbro. For once he was barely able to speak.

As part of the duties of both of his jobs, Outlaw is in the courthouse often, advocating and sometimes testifying for kids in trouble, or acting as a support for families of both victims and perpetrators. Sometimes he finds himself in the same courtroom where he was sentenced in 1989. The room hasn't changed in thirty years, and Outlaw runs into old associates: Patrick Clifford, who prosecuted him; his defense attorney, Ira Grudberg, still as colorful as ever; Tom Ullman, his second defense attorney, now promoted to chief public defender; and the bail bondsmen with whom he used to do business. But now there is an air of triumph to these encounters, on both sides. Outlaw is after all a success story, a fairly rare example of how the system can actually sometimes succeed. He has a particularly affable relationship with Clifford, who is now a judge. They smile upon seeing each other, joke around, talk about UConn basketball, the Celtics, the weather. Outlaw holds absolutely no grudges toward him. He calls him "a great guy." To Outlaw, Clifford is the proverbial "straight-up man," who was just doing his job in going after Outlaw in the courtroom all those years ago. A fair fight occurred between them, and Clifford won.

THESE DAYS OUTLAW lives in some ways a classic middle-class life. "I've been rich, and I've been poor, and I like being middle-class the best," he says. In the scant hours outside of work, his modest life revolves around visits with Pearl, who continues to be in and out of nursing homes; church with Germaine; and countless birthday parties and trips to water theme parks with his kids and grandchildren. (All told, he has six children by four different women, and thirteen grandchildren.) He recently took a weekend trip to the Basketball Hall of Fame in Massachusetts with a combined group of five children and four grandchildren under the age of five. Outlaw tries, not always successfully, to squeeze in parent-teacher conferences and trips to the mall for new sneakers. He saved up for months to buy a hospital-grade bed for Pearl to use in her apartment. He and Germaine put aside money from each paycheck for an entire year to save for a vacation to Puerto Rico. Outlaw posted a video from the San Juan beach. He is wearing a sombrero and sipping rum. Later he writes that "I never in my life thought I would be in love. She's helped me with my issues, made me a better man. . . . Germaine, you make my soul soar." He posts a picture of himself giving Joshua a bath, with the tagline: "Nothing better than this. Father love. Trying to give him what I never got."

But his life also teems with middle-class stressors. He returns to the townhouse he shares with Germaine at midnight, sometimes two in the morning, after being in the emergency room or at a crime scene, watches ESPN for half an hour, and tries to go to bed. No matter how much or little he sleeps, he's up at five-thirty the next morning. He likes to go to the gym but has no time. At one point Outlaw's son needed a car, and Outlaw gave him the white SUV that he had gotten from his old gangster friend. For months, Outlaw had to rely on rides to get around town, or simply walked. He wants to set a wedding date with Germaine, but she wants a ring, and he can't afford one. They attend weekly premarital counseling sessions. Outlaw knows he has a troubled track record with women, and he wants to get this relationship right. His cracked-screened iPhones ring constantly with calls from clients, coworkers, children, grandchildren. Often the call is from a former girlfriend asking that he pick up a child or come up with the copay for the latest medical appointment. By no means are all these conversations pleasant.

As if this were not enough, to make a little extra cash, he became a partner, with some former Jungle Boys, in a bodega that sells everything from groceries to rap CDs and DVDs. "All the necessities," Outlaw says without irony. (A clerk from the bodega will call Outlaw: "Yo, JB, do we have a copy of *Get Out*?" "Of course," Outlaw says. "In the front of the store, you'll see it.") He buys the goods for the store from wholesalers in New York, just as he once acquired cocaine in Queens. He opens the store at 6 A.M. on Saturday, his day off. It all seems like a kind of race—a slow and steady one—but a race nonetheless. Is it to atone for the sins of his past, to make up for lost time? It is not entirely clear. Outlaw perhaps does not know the full answer to that question himself: he just knows he must get to the next meeting, arrange the next trip to court with a client, facilitate the next gang mediation.

But the old life is not completely gone. Outlaw often wakes in the middle of the night. He sits up with a start, hearing the howls of people he stabbed in prison. In the bedroom, he wildly looks around for imaginary figures in the dark. In his mind's eye, he sees the dungeons of Lewisburg and Leavenworth, the smiles of Abdul Salaam and Frank Joyner, and dozens of the dead teenagers of New Haven collapsed onto sidewalks and streets, a blank, surprised look on their faces and the blood from their mouths running in pools. Outlaw gets out of bed, drinks lots of water, and paces downstairs back and forth across the living room to tire himself out, while Germaine sleeps undisturbed and oblivious upstairs.

But no matter what inner torments occur during the night, Outlaw is at the office the next morning, looking fresh, or mainly fresh, with a Dunkin' Donuts coffee in hand. He loves Goodwill but it's on the streets at night with the outreach team that he makes the most impact. He seems to breathe easier out on the street. In the old neighborhoods his pace quickens, and he has a fluency and an ease that he doesn't exhibit in the office, where he can feel hemmed in and uncomfortable with computers and copying machines. He has a particular impatience with office talk about weather and holidays. "After prison I lost interest in all that stuff. People say 'Merry Christmas' and 'Happy New Year,' and I don't really care. Rain, snow, or sun, it don't make no difference to me."

Outlaw feels the essence of both jobs is to be present and visible, every day. Jahad says that Outlaw is always the first to arrive at a shooting because he's already out in Newhallville or the Hill when the call comes in. To Outlaw, 90 percent of life is showing up. Hustlers got to hustle. "People in these neighborhoods can sense genuineness. And they can sense falseness. They are used to politicians only showing up during election time. When I was a drug dealer, I was out every day, blizzard or no blizzard. Same thing now."

And the homicide rate continues to go down in New Haven. The following chart, which exactly coincides with the years that Outlaw has been on the outreach team, shows the descent:

2010	23
2011	34
2012	17
2013	19
2014	12
2015	15
2016	13
2017	7

The numbers in 2017 represented a fifty-year low.

Meanwhile in Hartford, similar to New Haven in size, demographics, and post-industrial bleakness, the numbers are quite different:

2010	26
2011	27
2012	23
2013	23
2014	19
2015	32
2016	14
2017	29

These figures make Hartford, according to some measures, the eighth most murderous city in the country, ahead of even Chicago and Miami.

But then came the spring and summer of 2018 in New Haven. The headlines told the story:

APRIL 11, 2018: "New Haven Man Dies After Midday Shooting"

MAY 9, 2018: "Two Charged After New Haven Shooting"

JUNE 1, 2018: "Woman Found Murdered in Car in New Haven"

JUNE 23, 2018: "Man Killed in New Haven Shooting"

JUNE 24, 2018: "New Haven Homicide Investigated: 21-year-old man killed"

JUNE 25, 2018: "New Haven Police Investigating Shooting"

JULY 2, 2018: "New Haven Homicide Investigated"

JULY 3, 2018: "Triple New Haven Shooting Investigated"

By July there had been eight homicides in the city, one more than during the entire previous year. Not only was there more violence, but it seemed to be of a more unpredictable and unhinged nature, in a way that surprised and unsettled even the outreach team. Some of the deaths occurred on Sundays, a violation of the long-standing unwritten rule that the Lord's Day was off limits. The woman murdered in her car had been killed in front of her two children, a six- and an eleven-year-old. Jahad wrote on Facebook, "What kind of male shoots a woman??! In front of her children??? He has got to be brought to justice!" The outreach workers didn't even know some of the suspects in the killings, which was unprecedented in the history of the team. In some cases the perpetrators were recently released offenders who had been away for particularly long bids, or young kids—as Jahad put it—who had just "come off the porch." More than ever, crews, especially of the youngest kids, were based on murky social media connections. Jahad was forced to hire a college intern who spent hours on Facebook and Instagram trying to decipher the various shadowy interconnections. She uncovered that a group of eight boys, some of them barely pubescent, scattered through the greater New Haven area, had created a virtual gang, EBK, standing, apparently, for "everybody killers." On top of the surge of violence, the city made international headlines in August when ninety people overdosed one afternoon on the Green from K2, a synthetic form of marijuana laced with fentanyl. The event made New Haven ground zero in the national opioids epidemic.

Outlaw, and the entire team, were petrified about what might transpire over the remainder of July and into August. The heart of the summer, when murders historically hit their peak, was yet to come. Jahad said in the team's weekly staff meetings that everyone needed to step up their game. These weekly meetings took on the fervor of halftime in a losing football team's locker room.

Outlaw confided in Germaine that he would need extra support in the coming months, and would likely be working at all hours, with many long nights. Germaine responded with her characteristic magnanimity, saying she would help him in any way she could, including doing extra chores around the house, shopping and cleaning. But more than that, Germaine said, she would be there to listen and console, be a shoulder for him to lean on. Outlaw wept in gratitude, so thankful that he had met her. Outlaw found the time to buy extra bouquets of lilies.

Jahad said they needed to raise their game, but Outlaw already had. The homicide victim on April 11 was a burly, bearded man named Eric Lewis, shot in the chest and pronounced dead two hours later at Yale New Haven Hospital. Lewis, thirty-five, had a long record, including in the last year attempted carjacking, assaulting a police officer, and possession of PCP-laced marijuana. The killing appeared to be gang-related. Word of the shooting spread on social media, and soon, a hundred people were angrily milling around the sidewalks outside the emergency room. Outlaw, as always, was the first member of the team to arrive at the scene. It was an early spring day, fifty degrees outside, with pale sunlight. The trees were stark, not a leaf on them. Time began to slow down for Outlaw, the way it always did in such heightened circumstances. Before taking any action, he surveyed the crowd, looking for indicators of danger, just as a police officer would.

Pepe was the second outreach worker to arrive. Outlaw told him that he had already "peeped"—that is, analyzed—the scene, and briefed him on the situation. Eric's girlfriend was screaming on the sidewalk. The police had impounded her car, presumably for evidence, and she seemed hysterical about what they might find in it. With a tilt of his chin, Outlaw drew Pepe's attention to a group of men whom Outlaw knew to be aligned with Eric and who appeared to be hungry for revenge. Outlaw whispered to Pepe that he was concerned about one of the men in particular, who was especially agitated. Outlaw had seen the man park his car illegally in a nearby spot where he could make a quick getaway if necessary. "Let's go talk to him," Outlaw said. Pepe nodded, and they approached the man and gave him a "bro hug," a quick half embrace. As Pepe wrapped his arm around Eric's associate, he subtly felt around the man's belt to see if he was armed. Releasing the man, Pepe gave Outlaw a nod that indicated the man was carrying a gun.

Outlaw addressed the man directly: "I know why you are here. You know why you are here. We get it. We know you angry. But man, you gotta go. Nothing good is going to happen out of this." The man, clearly disturbed, was barely able to stand still, but he listened to Outlaw. Even in the midst of his fury, he seemed somehow relieved that Outlaw told him it was okay to leave. It was well known on the streets of New Haven that if William Outlaw told you to do something, even if it could be perceived as a retreat, you would still be "okay": you would still have street cred. Outlaw's word could give anyone the necessary cover. All one had to say was, "Juneboy told me to do it." With a combination of reluctance and relief, the man walked to his nearby car and drove away. The police apprehended the suspect in Eric Lewis's murder in North Carolina a few weeks later.

Over the summer, Jahad instructed the team to carry out specific tasks, based on their individual strengths and expertise. Doug Bethea generally did not go to crime scenes because it retriggered the trauma of his son's death, but he often conducted home visits. He went to the apartment of one of the boys who had been shot in the spring. They sat on the porch of the boy's house in Newhallville. The kid had just been released from the hospital and showed his leg wounds to Doug as if he were proud of them. The boy told Doug that he had to retaliate for the shooting, to show that he was tough, to save face. Doug heard him out, without judgment. Then he told him about Scotty. "At your age you may think you're immune but you're not. It can happen in a second. No one is immune, let me tell you," he said. The boy never did take his revenge.

JULY 4 WAS the hottest day of the summer, with a heat index of 108. The outreach team, along with people from many other organizations, participated in a basketball tournament in Newhallville. Also present were the police, many of whom wore street clothes and handed out cold drinks. The kids were shocked but appreciative of this gesture, except for one young tough who told them, "I ain't gonna accept nothing from you motherfuckers." Jahad wrote on Facebook that night:

> Great job today by the Newhallville Neighborhood Corporation at the basketball tournament at Bassett Park. Much appreciation to Beulah Heights Church who kept their promise and attended the games—they kept their word, now we need more churches involved and engaged. We will keep pressing forward. . . . Lastly, thanks to Lt. Karl Jacobsen and NHPD for supporting the event and buying cases of iced bottled water for the players and freeze pops for the children. I hope to see everyone next week and support the games.

The next day, Outlaw posted on Facebook:

What Can You Do !!!!!! To Stop The Violence In Our Community ???????

His challenge drew twenty-five comments, most of which had to do with prayer and the black community needing to heal itself. Outlaw and Jahad did not disagree with these opinions.

Then Jahad found the money for, and created for the first time in the team's history, a summer camp program, a four-day-a-week, five-week program. Jahad could have enrolled thirty kids, but he only had the funding for eleven. The boys were paid $10 an hour to clean up parks, and also to play basketball and card games and read poetry by Maya Angelou and Langston Hughes. Jahad read to them these lines by Angelou:

> I do my best because I'm counting on you counting on me.
> Do the best you can until you know better. Then when you know better, do
> better.
> Develop enough courage so that you can stand up for yourself and then
> stand up for somebody else.

Some of the boys didn't connect with the writing and the poetry. "Think of the writing like rap lyrics," Jahad said. "And then write down for me what you think *your* best self is." Jahad held weekly competitions to see which members of the group could memorize the most lines from Angelou and Langston Hughes. The boys often performed the lines in rap fashion.

The camp met each morning at nine-thirty in the community room at the New Haven Family Alliance. At the first meeting, Jahad and Outlaw asked the boys who had lost a sibling or a parent to violence. Five of the eleven boys raised their hands. The shared trauma created a bond among the boys, as did the fact that seven of them played varsity sports, mainly football. The kids set up a GroupMe to stay in touch with each other.

Toward the end of the summer, Jahad and Outlaw took all eleven of the boys for a tour of the Cheshire prison, where Outlaw had finished out the last years of his sentence. The "campers" were given a tour of the cells, the chow hall, the gym, the workshop where inmates were paid $20 a month to make license plates. In the auditorium afterward, they attended a lifers group, a support group for men who were serving life sentences. When Outlaw walked into the room, he realized that he knew three of the men and had started his prison tenure with them in the late 1980s. He had run with them on the streets and perpetrated similar crimes. Outlaw's legs almost buckled. He embraced them, sobbing. He tried to speak for some time but his lips could not form words. By all rights, if not for the grace of God, he would be in that lifers group now. When he finally could speak, all he could say was "I am no different from you all."

Jahad asked the boys to introduce themselves. When the kids spoke, none of them said what neighborhood they were from. They all simply said they were from New Haven. Outlaw and Jahad smiled at each other: they had emphasized all summer that the kids should not think territorially. Outlaw and Jahad had repeated to the kids all summer: "No, no, no! We don't deal with geography! You all from New Haven! Every hood is our hood!" Outlaw recalled that during the first meeting of the camp the boys had introduced themselves by naming the blocks they lived on. The older men of the lifers group, each in their own way, said to the kids that they deeply regretted their lives, that they suffered from depression and agitation, and that they constantly relived the crimes they had committed. Their only solace was the brotherhood of the group. One of the boys—whose brother had been killed the year before—asked to speak. He talked openly of his depression, and said, "If it wasn't for Juneboy and Jahad, I would have committed homicide or suicide a while ago. I was under a lot of pressure for revenge. Now I have some hope, some room for cover."

On the way back to New Haven, Jahad and Outlaw took the boys to an all-you-can-eat buffet. Many of boys, whip thin, were clearly used to starving. "Eat all you can," Jahad said. "Eat up the whole damn place."

In New Haven, there were no homicides in the remaining days of July, August, September, or October. Meanwhile Hartford recorded its eighteenth homicide in November.

WHAT HAS OCCURRED in New Haven over the last eight years represents an unprecedented turnaround, and it has been duly recognized on the national stage. On three occasions in 2015 and 2016, President Obama invited New Haven's mayor as well as its then police chief, Dean Esserman, to Washington, D.C. Esserman, based on his successes in the city, advised a White House task force on crime control. "I heard good things about New Haven," the president said to him in a private moment. The *New York Times* published an editorial touting Connecticut's success: "Overall, crime in Connecticut is at a 48-year low and falling faster than almost anywhere in the country." New Haven's reduction was by far the largest of any city in Connecticut.

Two things have been widely publicized as the cause of the miracle on the streets of New Haven. On the face of it, Chief Esserman brought a more inclusive approach to policing than his predecessors. He got significantly more officers to walk neighborhood beats, and reenlisted the community's help in sharing information and solving crimes. But in late 2016 Esserman was placed on extended medical leave after a history of complaints regarding his personal conduct. Eventually he resigned and was replaced by Anthony Campbell, a multidecade veteran of the force. Campbell's mother had been a prison guard at Rikers, where she met his father, who was incarcerated for drug dealing. Campbell went to Yale College, also receiving a master's degree in divinity at the university, and then became a beat cop. Campbell, who is African American, is highly respected and has presided over a renaissance in the department.

The other widely publicized factor for the turnaround in New Haven has been a federal anti-gang initiative with the name of Project Longevity. Project Longevity has received a lot of positive press and was extolled by James Comey, then director of the FBI, on a visit to New Haven. The initiative is ably run by the former Newhallville beat cop Stacy Spell, who spent the early part of his career trying to bust William Outlaw and whose life Outlaw may have saved in the late 1980s. Project Longevity is a kind of "scared straight" program for gang members. High-risk individuals are forcibly "called up" by parole officers to a meeting at the downtown community college, where they are implored to turn themselves in or otherwise cooperate with police, prosecutors, and members of the public. In exchange, they are offered social services and potentially reduced sentences. At one meeting, twenty presumably gang-involved men, all young African Americans, sat in a circle, surrounded by a crowd of one hundred mostly middle-class and upper-class white professionals. Spell presided over the meeting with an engaging magnanimity, but it did not dispel the ill will that seemed to emanate from many of the "called up" men. Yale researchers have shown that Project Longevity has been effective in reducing short-term violence, by getting the highest-risk individuals off the streets, but even its developers admit that Project Longevity is not designed to effect lasting change.

Certainly, the new leadership in the police department and Stacy Spell's efforts at Project Longevity had played their role, but to many people, William Outlaw is the biggest reason for the turnaround in New Haven.

"WE KNOW THERE'S a contagion factor around youth violence," says Dr. Patrick Hynes, a sociologist and director of best practices of the Connecticut Department of Correction, who has followed Outlaw's career. "One incident leads to another. If you stop the initial event, you might stop six further ones. In thirty years in the field, I can't think of anybody who understands the dynamics behind these conflicts better than Outlaw. He's good at sizing people up. He's intuitive and authentic. He was highly competent as a gangleader. And now he's highly competent as a gang interrupter. But it's all the same skills at play. I think there's something deeply heroic about what he's doing."

Ivan Kuzyk, the chief analyst for crime statistics for the State of Connecticut, knows more about the contours and trends of crime in Connecticut than anyone. Kuzyk says: "In New Haven eight years ago there were over thirty murders. The vast majority of the victims were young African-American men from a few neighborhoods. Afterward the community got together and said we cannot stand for this anymore. Distrustful as they are of the police, they resolved to create a community solution to the epidemic, as opposed to a law enforcement one. But it's only a figure like Outlaw that can be the catalyst for that kind of change. Typical social service programs can't do it, we in the state sector can't do it, and certainly the police can't do it. With that body and his size and ability to understand the nuances, he changes the dynamic. He's what they said about Steve Jobs—he's a 'first arriver.' William Outlaw is Bill Gates and Al Capone put together."

Some people, including Spell himself, think that Outlaw and the outreach team are responsible for two-thirds of the crime reduction in New Haven. "Juneboy and his crew have turned things around more than any group in town," Spell says from his tenth-floor office in the federal building in downtown New Haven. Wearing a two-piece suit with a gun holster in his belt, Spell says about Outlaw, "He is the ultimate alpha male," a remark that applies equally to himself. "With that size and those smarts, who's not going to listen to him? I would!" he exclaims, bursting into laughter.

"This is the story of a phoenix rising," Spell continues. "This is a one-in-a-million character, a one-in-a-million talent. I remember running into Juneboy on the streets after he was released. 'Hey, Big Stace, how are you?' he said, and gave me a big man-hug. I could tell he was the same guy that he had been before he went in, that he was still all there. To survive Lewisburg with your faculties intact? To survive Leavenworth without joining a gang? Who even does that? *Nobody* does that."

"Outlaw has what I call a lightning mind," says Mike Lajoie. "I've only seen it a few times in my career, and only among the greatest criminals. They can see things four steps ahead of where you can, like a chess master. They can't necessarily articulate it, but they can feel it. He's now using those same skills. He is changing the world."

Ivan Kuzyk says, "If William Outlaw had been born in a different place, to different parents, in a different time, he would have become a Fortune 500 CEO. As it was, he took mediocre talent and created a first-class gang that ran half of the city of New Haven. What he accomplished was the equivalent of the Afghan warlords putting together scrubs and taking on the U.S. Army."

Kim Goodman, still a parole officer in New Haven, says, "I think what Outlaw has done is not really quantifiable. It is more of a feeling. How mentally strong do you have to be to survive a year in solitary? And to come to terms with his horrible past? I think he offers an example of how one individual can provide to the community the maintenance of hope."

"Have you seen him on the New Haven Green?" says his Goodwill colleague Mary Loftus. "People come out of nowhere to be around him. Homeless people, hot dog vendors, policemen, Yale professors, ex-convicts, you name it. I call him the 'Governor of New Haven.'"

Ed Kendall, Outlaw's former police nemesis and now colleague, puts it in a more succinct and street-smart manner. "If Outlaw can change, anyone can."

THE MEN SHUFFLED into Outlaw's weekly Men Helping Men therapy group for recent parolees in the large conference room in the halfway house as if they were still in prison, walking almost single file, keeping to themselves as much as possible. It was a September day, but it still felt like summer outside, humid and warm. The air conditioner made a loud humming sound. Out of the third-floor window, the participants had a view of the cramped neighborhood: a bodega across the street, as well as a check-cashing place, a funeral home, a homeless shelter, a number of liquor stores, and a lawyer's office advertising "We speak Spanish/*Se Habla Español*." A few weeks later there would be a shooting outside the shelter, and a person would be seriously hurt, but the story wouldn't even make the local news.

In the back of the room was a table with lasagna, salad, and garlic bread. The men waited in line, served themselves, and sat morosely at a large rectangular table. They ate, some ravenously. For some men, it was obvious, this was the only meal of the day. Most tried not to make eye contact with their neighbors if they could help it.

The talk before the group was that one of their members, Jason, had been "violated" in the last week. "Violated" meant sent back to prison. The update was delivered by a very slight, very young man with cornrows and a drawn face. He couldn't have weighed more than 130 pounds. He delivered the news about Jason with barely constrained excitement, explaining that the parole officers had come into their halfway house with dogs who had detected cocaine in Jason's room. Jason was immediately cuffed up and removed. The men listened to this story but didn't say much.

Outlaw entered the room, wearing all black. The gold in his teeth seemed brighter than usual, in contrast to his clothes. Outlaw was sweating and breathing heavily, having just clambered up the two flights of stairs. He was late, having come from prison. When he heard the news about Jason, he said, "I'm not surprised. I saw it in his eyes last week. I thought he'd been using cocaine." The food finished, the plates cleared, the men looked expectantly at Outlaw. The mood of the room was quiet, with just the sound of the air conditioner rattling away in the background.

Outlaw handed out the weekly "Job Leads" flyer. He had entitled the sheet "Snag-A-Job":

Party City Fulltime Merchandise Specialist

Kentucky Fried Chicken Hamden Team Member Full/Part Time

Wendy's 2195 Dixwell Ave. Hamden Crew Member Full/Part Time

Laquinta Inn & Suites 400 Sargent Drive, New Haven Maintenance/
       Laundry Attendant Full/Part Time

"Get out there. No excuses!" Outlaw said. Then, changing pace, he said, "We're going to do an exercise tonight. It's deep, real deep." From a leather briefcase he pulled out a small rectangular mirror with green plastic around its edges, as well as a stack of yellow paper and twenty carefully sharpened pencils. Outlaw held the mirror up and peered into it, gazing at his face. He smiled at himself, admiring his image for comic effect. The men laughed.

"Normally when you look in the mirror you just look real quick, you know what I'm saying?" Outlaw said. "If you're a dude, you probably just do it once a day, in the morning. Check your hair, see you don't have nothing in your teeth. But this exercise ain't nothing like that." He passed out the sheets of paper and the pencils.

On the sheet of paper was written:

THE MAN IN THE MIRROR

Things I like about myself   Things I don't like about myself

_____   _____

_____   _____

_____   _____

Additional things I want to say about myself

_____

"What I want you to do," Outlaw said, "is hold this mirror up to your face for two minutes. I will time it. Before you hold up the mirror, I want you to say, 'Who am I?' I first did this in a group in the Feds. I learned a lot about myself from this exercise." Outlaw held the mirror in front of him and said, loudly, "Who am I?" He looked at the mirror directly, not quite for the full two minutes, but for more than a minute. The room was silent, the men watching attentively.

"What happens is that you just don't look at your hair, or teeth, or whatever. You see yourself, just you. And some people can't handle that. Then after you do the exercise, I want you to write on the paper about what you saw, the shit you like and the shit you don't." Outlaw passed the mirror to the very young man who had told the story about Jason. He nervously held up the mirror to his face, and gazed into it.

"Who am I?" he said unsteadily. The young man began with a smile, as if there were something vaguely funny about the exercise. But the smile quickly dissipated. Thirty seconds into the exercise, his hands shook. He shut his eyes and his whole body started to tremble. A distraught look came over his face. Then the young man abruptly put down the mirror, got up, and left the room. An embarrassed silence followed but it didn't last long. The man next to him, a studious-looking, bespectacled African-American man wearing a collared shirt, picked up the mirror and said more confidently, "Who am I?" and stared into it for two minutes. Then he passed the mirror to his neighbor, and began to write on the piece of paper. And it went on like that. One man cried. Many of the men's hands trembled as they stared into the mirror. One man laughed. Others were expressionless. All of them wrote assiduously on the paper after they had finished looking at themselves.

At some point the young man returned to the room and completed the exercise. Filling out the paper, he handled the pencil as if it were a suspicious object, something that he was unused to. When the last man had finished, the mirror was returned to Outlaw, who said, "Now I want all of you to read to the group what you wrote on the paper.

"You first," he said to the young man. He read, nervously:

Honest	Crack baby
Loyal	Little boy inside a man
Giving	P.O.M.E.

"P.O.M.E.—that means product of my environment," the young man explained to the group.

"Yes, it does," said Outlaw, "and that's something that I know about. . . . Look, brother. I want to thank you for your honesty. Thanks for sharing. And thanks for coming back into the room. That was brave of you. It's the longest two minutes ever, right?"

"Yeah, it is. I don't like what I've been," the young man said. "I blame my mother and father. And now I'm stuck with the man in the mirror."

His neighbor, the one with spectacles and the collared shirt, read:

Good-hearted	Evil if I have to be
Caring	Can't turn off feelings
Reliable	Hard time forgiving
I struggle with depression	I see someone who looks tired

"Thank you, brother," Outlaw said. "You're very brave. I have a lot of respect for you."

A tall man went next. He was in his twenties, gangly with tight corn-rows. He couldn't sit still. He read:

I still care	Low patience
Staying Strong	Anger
Leader	Still have criminal thinking
Still a young black man	I still like to play around

Outlaw said, "I know how that is . . . I know about being a young male! I do, I really do. I know what that's like. I sowed some wild oats. I did my fair share of that, know what I'm saying?" The group laughed. "But all that anger creates anger. And you know what? What I always say is that the only people you can't fake out are you and God."

One of the men, the oldest, in his fifties, the one who cried during the exercise, read next. Outlaw asked him, "Why were you crying?"

Spanish was the man's first language, and he spoke in accented English. "Reason I cry is I have job now. Last time I do this exercise I don't have no job."

"God bless you, brother," someone said.

The group was winding down. It was now dark outside. An ambulance siren sounded somewhere in the distance.

"Listen up," Outlaw said. "Some of us come to this group because we have to. Some of us come for the food." Laughter from the group. "But for me this is my therapy. This is where I get all my emotions out, all of the stress. This group is real for me, man. This is the highlight of my week."

The men nodded with a surprising kind of earnestness.

"One more thing," Outlaw said, as the men were just about to leave. "A psychologist I saw at Lewisburg said, 'William, if you don't get ahold of that anger inside you, you're always going to be a product of your environment.' I was angry, really fucking angry, man. Furious at my father, at my grandfather, whatever. The psychologist said, 'I understand that, but there's really nothing I can do for you until you get ahold of that anger.'

"And that's when I learned to let it all go."

ON ANOTHER EVENING, Outlaw was driving in his white SUV in New-hallville. He had just met with some kids in a park, telling them about the outreach team. Outlaw's cell phone rang. At first, it was hard to tell who was talking on the other end. Eventually, Outlaw realized it was Mary Loftus, calling from her evening job at a homeless shelter.

"Juneboy, this woman I work with at the shelter, Lisa Craggett . . . you know her? She's almost peeing in her pants. She's hysterical." Of course Outlaw knew Lisa Craggett. He hadn't seen her in thirty years, but he had run with her in the streets back in the day. Mary continued, "Lisa just got a phone call. A friend of hers is saying that she thinks her boys were shot in the Hill. Corner of Davenport and Vernon. . . . She's trying to get a ride down there now. Can you get there?"

Within minutes, Outlaw was there. Dozens of people stood around two wrecked cars in the middle of the intersection. A red Chevy sedan with bullet holes in the side had collided with a white van. There were two ambulances and three police cars and a number of officers cordoning off the area. Outlaw approached one of the officers, who he knew was sympathetic to the outreach team. The officer said that one or two masked men had ambushed the red Chevy and fired at the boys in the car. All the victims—four of them, it seemed—had been taken to the emergency department at Yale New Haven.

Outlaw went to the hospital, one block away. He was greeted by a police officer who waved him inside. The Yale New Haven Emergency Department was newly refurbished, sleek and beautiful, with floor-to-ceiling windows affording views of tree-lined York Street. But inside was a scene of torment. Twenty or thirty people were milling about, all in various levels of distress. Outlaw approached a man whom he knew to be Lisa's brother-in-law. The brother-in-law confirmed that Lisa Craggett's two sons, Jacob, age fifteen, and Joshua, age twenty-three, were being treated by doctors, as was the driver. The fourth person in the car was unscathed. The brother-in-law confirmed the four had been ambushed at the corner by two men, guns blazing, like something out of the movies. Outlaw didn't know Jacob, but he knew the older brother.

The only member of the family who wasn't there was Lisa Craggett. The brother-in-law said she was still at the shelter, too frantic to drive.

Outlaw said, "You want me to pick her up?" The brother-in-law said that no, he would go, and he'd be back in half an hour. In the meantime, Outlaw called in members of the outreach team. It was 10 P.M. now, and an officer named Sergeant McCarthy asked to speak to Outlaw alone. He took Outlaw aside in a hallway off the waiting area. "Look, Juneboy, one of Lisa's sons died in the back. I don't know which one." Outlaw's stomach seized, but he didn't have long to process this news. Lisa Craggett had just walked in, hyperventilating. The extended family surrounded her. Outlaw stayed back, but Lisa saw him and approached him. Outlaw hugged her, feeling the agony of knowing what she didn't know.

In the waiting room, groups of family members prayed, while the stepfather of the boys was saying he was going to get revenge on whoever shot his sons. Outlaw told him, as calmly as he could, to let the police do their job, that they knew what they were doing, and that they would do their best to capture the shooters.

A doctor came forward. He asked to meet with Lisa and the boys' stepfather in a private meeting room. On their way, Lisa said, "Juneboy, come with us." In the conference room, after they all sat down, the doctor said, "Mrs. Craggett, I have very sad news. One of your boys did not make it. Your older son. He passed away approximately one hour ago." Lisa shouted, "Bring me to him!" The doctor rose and brought Lisa to the treatment rooms in the back. Outlaw was left alone in the room. He was having a hard time breathing. He eventually went back to the waiting room, where Lisa returned ten minutes later, barely able to support herself.

"Those fucking doctors. They got the wrong son! It was my baby who died! They killed my baby! They killed my baby!"

Jacob, the fifteen-year-old, was six foot three, 270 pounds, and much bigger than his older brother. The doctors had just assumed that it was the older son who had died. Lisa was apoplectic; no one could approach her. After a while, Lisa's pastor was brought in, and he was able to take her and the family aside in an attempt to console and counsel them. Occasionally as the night wore on, Outlaw went out to York Street to get air. Forty kids stood out there. A number of them were football players: Jacob had been a star player on the Hillhouse team. The football players were mouthing off about hurting whoever was responsible. Outlaw said, "Let the police do their job. Don't do nothing stupid."

Hours passed. Finally, at five A.M., one of the doctors came out to talk to the family. They had good news: the two other wounded boys were out of critical condition. They were going to survive. The crowd began to disperse. Outlaw stayed for a while, but there wasn't much left for him to do. He told the outreach team to go home.

When he left the hospital, Outlaw was exhausted and in a state of profound numbness from the night's events, but he knew exactly where he needed to go from there. He drove on Interstate 91 a mile and a half north, and pulled into a parking lot, fifty yards from where he grew up at Q View. The wide river lay in front of him, surrounded on all sides by the highway and apartment buildings. Outlaw got out of the car and looked out over the water. The sun had just now come up. He walked to the edge of the river. The interstate was fifty yards away, making a constant, steady drone from the tractor trailers and cars even at this time in the morning. Outlaw leaned down by the riverbank and picked out some flat rocks. He skipped about fifteen of them across the water. It was satisfying to see the rocks skim over the river. Then he walked toward the highway. Outlaw sat down on a slab of concrete, under the highway, where it was cool and dark. He shut his eyes.

Jacob was the fourth young person killed in New Haven in the last month. The scenes from the dreadful night filtered through his mind like a horror movie. The wrecked car. The police. Lisa. The phone calls. The doctor. The screams. The faces of the kids outside the hospital . . . Outlaw started sobbing. During the night, he'd been doing his job, trying to keep people calm, praying, getting them water, talking to his team, talking to the doctors, keeping his eye on the kids. But now it all hit him. After a while the images settled down but what he couldn't get out of his head was the screaming, the sounds of the family wailing. The sound of Lisa, hysterical. It wouldn't stop in his head: *My baby died. They killed my baby!* In time Lisa would become completely disabled from the trauma, unable to work or leave her apartment, where she built a shrine to her son. But the main thing that Outlaw thought about, and which he couldn't get out of his head, was his own involvement in similar tragedies. *I caused that kind of pain to other families. That same crying that family had all night—I caused that—I did that to other people.*

He lay there for a long time, but after a while, he knew he had to get up. He said a prayer of gratitude that he was even alive, that he was breathing, that he had the opportunity to be at the hospital last night and to look out on the Quinnipiac River this morning. The main thing, though, was that he needed to get busy. There might be retaliation after Jacob's murder, and he had to be out on the streets. He got into his SUV and drove to meet with some kids in a park. It felt good to be moving again. Hustlers got to hustle.

ABOUT THIS PROJECT

Charles Barber and William Outlaw, Wesleyan University, 2017.
Alex Fuller, Cianci Creative, LLC.

I FIRST ENCOUNTERED WILLIAM OUTLAW ON A PALE MARCH DAY IN NEW Haven in 2014. We met on the second floor of a church at the end of a highway exit ramp in an industrial area of New Haven. It was the end of an interminable winter. As I parked the car, there were still six inches of crusty snow on the ground. The auto parts store across the parking lot displayed the sign: IF YOU WANT TO FIND YOUR TOOLS WHEN YOU COME BACK TO YOUR TRUCK, LOCK YOUR DOORS.

I walked into the lobby, where a sleepy guard asked me to sign in. It was two o'clock in the afternoon and I was the third visitor that day.

The church, called the Church on the Rock, was on the ground floor. Various social service agencies were upstairs, where I was scheduled to meet William. I walked down a long windowless hallway to the offices of Goodwill. I was buzzed through two automatic aluminum doors and greeted by a secretary.

"I'm here for William Outlaw," I said.

"Juneboy?" she said.

She led me to a conference room while she looked for Juneboy. I sat at a fake-wood conference table and looked out at gray New Haven: a housing project, numerous empty lots, and a ceramic-tile distributor with a barbed-wire fence around it.

The meeting had been arranged by Ivan Kuzyk, the director of the Statistical Analysis Center for the State of Connecticut, which tracks criminal justice trends. Kuzyk had seen Outlaw speak at a legislative committee on behalf of employment opportunities for ex-offenders and became interested in the impact that he was having in New Haven. I had known Ivan for years, and he knew my writing. Outlaw was looking for a writer to tell his story, and Ivan thought we might be a match. I remember Ivan's words exactly: "He has a big-time story."

A few minutes later Outlaw entered the room. The thing that struck me first was his enormous size: six feet four inches tall, and—as he told me later—410 pounds. His arms were the circumference of an average person's legs. His hair was styled in a closely cropped Mohawk, and his face was largely unlined, to the degree that he could pass for ten years younger than his forty-six years. He wore a red rugby shirt, blue jeans, and Timberland work boots.

"How ya doing," Outlaw said. It was less a question than a statement—the universal male way of greeting another male. He spoke slowly, in a gravelly voice.

I noticed, as he spoke, a flash of gold from his front teeth. These, I was to soon learn, were the vestiges of his former life, a gold *J* engraved on one incisor, and a *B* on the other, the letters now faded beyond recognition.

He began talking a little about his story. He was extremely matter-of-fact as he spoke. He talked about Q View, Pearl, the Jungle, a series of stays in federal prisons all beginning with the letter *L* that I couldn't keep track of, and about coming back to New Haven and starting over. He didn't elaborate much. He said if we were going to work together, he didn't want "no hood book." "The thing about me," he said, "I don't like to talk about war stories. I'm more interested in the change part." He wanted a book that the kids on the street would read. It seemed that he wasn't particularly proud of the Jungle Boys, and the way he spoke about the gang seemed to represent an accumulation of a lot of hard times. It reminded me of the way my father spoke about being in combat in World War II. I remember being chilled by something that Outlaw said. He talked about eating a cheeseburger at a restaurant after killing Sterling Williams. It made my stomach turn.

I asked Outlaw at that first interview why he changed. He told me the story about Frank James instructing him that he needed to make something of his life—if not for himself, then at least for Frank. I thought I detected tears in his eyes. (Much later, when I read William the passages I wrote about Frank and Warren Kimbro, he cried openly.) We talked easily for an hour. I told him a little bit about me, but not much. I wasn't sure I trusted him. When he talked about Long Lane, I didn't tell him that I had grown up across the street. (Which is to say that when Outlaw was 12 and I was 17, we lived 400 yards from each other.) As we walked out after the interview, Outlaw told me with what I was to learn was a characteristic decisiveness, "I'm ready to tell you my story."

The problem was that I wasn't sure I was ready to hear his story. I was in the middle of writing a novel, and I was enormously busy with writing and teaching, but more than anything, I wasn't quite sure I wanted to enter Outlaw's world, a world of apparent redemption but also of unremitting violence. I also felt this huge abyss between us, an enormous cultural and racial space that felt fairly insurmountable. Although I do not come from wealth, my education can be described as the epitome of white privilege: prep schools and Ivy League universities. The most criminal thing I've ever done is to get a speeding ticket. I have an obsessive, guilt-oriented nature and fret at the slightest thing I think I might have done wrong—and even things I didn't do. (Outlaw soon enough picked up on my nature. Sometime later—after I learned that he was an expert evaluator of talent—I asked him if I had been in the Jungle Boys, what he would have had me do. ("I'd have you in the back counting the money," he said, laughing.)

The other reason I was cautious was that shortly after meeting him, I was warned not to write about William Outlaw. As I was feeling out the project, I met with the then New Haven police chief, who refused to talk to me about Outlaw. He said, "Every police chief has stories he is unwilling to share," and accused me of being impossibly naive to even entertain an interest in writing about Outlaw. Virtually everybody I met in the Connecticut law enforcement community knew about William Outlaw and his family. A lawyer explained, "In every big city, there are one or two families that are notorious for criminal activity. It is like it is passed down through the generations. In Hartford, it's the Hightowers. In New Haven, it is the Outlaws. There are generations of Outlaws, and all of them are criminals."

Not long after I met Outlaw, I was talking with a friend, a court administrator in the state judicial system. I asked him if he knew of William Outlaw. Of course he did. He was deeply skeptical about Outlaw, about his current activities, and anything at all to do with him.

"What about all the good stuff he's doing in New Haven?" I said. "It looks like he's saving people's lives."

"Listen," my friend said. "When I was young and just starting out at the court, a judge took me aside. He said: 'There are three types of people in the criminal justice world. There are citizens, people like you and me, who do something stupid and get caught, or don't have a good lawyer. They're not fundamentally criminal. Second are the mopes. Mopes are people with mental illness, substance abuse, often below-average IQs. They're not really criminal either—they just get caught up in stuff. And finally, there are the bad guys. They're in the minority, but they're really, really bad. They are the only true criminals in the system, and they influence everyone else."

My friend looked at me. "I don't care what he's doing in New Haven now. Outlaw is a bad guy. And bad guys never change."

—

Despite the warnings, William and I began spending time together. In the beginning I gave him rides because he had given his SUV to his son. For some time he didn't have a car. (At one point he asked to borrow money from me to buy a used car, but I refused, saying I didn't want to alter our professional relationship.) We went to the Jungle, to Newhallville, to the Hill, to Q View, to his bunking spot under the bridge. I had worked in New Haven for years, but I didn't know any of the neighborhoods. My New Haven was essentially Yale or the central area of downtown, concert halls where I'd seen Miles Davis and Marvin Gaye and the Talking Heads, and lecture halls where I'd seen James Dickey and Robert Penn Warren. I had never been to Newhallville, half a mile from campus. Nor did Outlaw really know my areas of town. Sometimes we went to cafés that I knew, near Yale. He was much quieter in these kinds of places than he usually was.

Notwithstanding our differences, we developed a rapport and turned out to have a number of things in common. We are both "big picture" people, prone to strategic thinking and analysis, and shared various arcane interests. We both had the misfortune of being New York Jets fans, which became a source of shared commiseration, and we talked at length about boxing history, how Ali outfoxed George Foreman in Africa, and about long-ago fighters like Jack Johnson or Sugar Ray Robinson. These were conversations I could never have with my academic colleagues. Out on the streets, I saw Outlaw's magnetic way of talking to kids. I saw his strategies unfold before my eyes, how he would be confrontational with the swaggering kids, and so soft-spoken with the vulnerable ones that it was difficult to make out what he was saying. I saw people throng to him, and the way his coworkers treated him with respect. I witnessed him working sixteen-hour days without complaint, although he would nod off very occasionally after a late night in the emergency room, saying, "Sorry, man, I'm tired."

At one point, after having been in regular contact for about five months, I was unable to reach him for about two weeks, which was unusual. I left numerous messages for him to call me. I wondered if he was no longer interested in the project. I remember finally reaching him on a blindingly bright August morning. It was eleven A.M. on a Sunday. He picked up the phone with a whispery croak. He sounded ill. "Are you sick?" I asked. "Not . . . not really," he stammered, barely able to speak. He explained that four young men had been murdered in the last two weeks (one of these was Jacob Craggett). He said he was in a dark bedroom with the shades down and hadn't been able to get out of bed for two days. "It's too much, Charlie. It's too much. I can't do this no more," he said.

That was the moment I decided to write about him.

FROM THEN ON, we met virtually every week for the next four years. I conducted hundreds of interviews, and spent hours with him out on the street, at meetings of the street outreach team, at his Men Helping Men group, out in the neighborhoods. (I gave up taking him to the Yale cafés.) It turned out that we had oddly parallel lives. We were both overworked, and had mothers who were in and out of the hospital. Often we had to cancel on short notice: in my case for grant and writing deadlines; in his case, shootings and ER visits. Neither of us got annoyed at each other about the schedule changes. We developed an affable, understanding relationship. I arranged for William to speak at Wesleyan University about his story, during which he said, "I ain't opened up to nobody in my life more than Charlie, even more than my fiancée." Germaine was in the audience, and smiled with characteristic grace. Together we obtained three hundred pages of prison records and court documents, and with William's full cooperation (often he set up the interviews with a simple "Yo, my writer wants to talk to you"), I interviewed corrections officers, judges, attorneys, former girlfriends, therapists, William's colleagues, and former gang members. Outlaw would leave the room when I met with others. "I don't want to interfere," he'd say. I found a remarkably high correspondence between Outlaw's stories and those told by witnesses to the same events.

But I understood the skepticism about Outlaw. I had been immersed in criminal justice issues for almost a decade before I met him and knew that the research indicated a person like Outlaw was unlikely to change. After working for many years in the mental health field, in 2004 I fell into a job as a trainer and consultant to the Connecticut Department of Correction. What I learned was sobering: most of the treatment interventions to reduce recidivism that had been employed historically did not work and in fact made criminals more likely to recommit crimes. The following had all been shown to be ineffective: the experience of incarceration itself, which boosted one's chance of committing or recommitting future crimes, Scared Straight programs, wilderness-encounter programs, boot camps, psychotherapy, and programs designed to boost self-esteem. The research showed that only two things were consistently effective—treatment for substance abuse, and a specialized form of cognitive behavioral therapy that helped offenders address the distorted and criminal nature of their thinking and taught them new social skills.

As part of my job, I conducted one hundred interviews of prisoners at Osborne, where Outlaw had been imprisoned. In fact he would have been incarcerated there (we never met) while I conducted risk assessments, which are psychological questionnaires intended to predict a person's likelihood of reoffending. I associate those first visits to Osborne with indelible images: dozens of prisoners, of all shapes, sizes, and colors (although more than half were black), in the same beige uniforms walking single file down a yellow-painted median in the hallway; the extraordinary amount of downtime the prisoners had; touring a well-equipped woodshop that had been closed for lack of funding (this would have been where Tiny Piskorski hung out); the interminable processing time it took for me to get into the facility. (Finally, I asked a colleague why the corrections officers were taking so long to let me in. "That's easy!" he said. "They think you're a defense attorney.") I found that many of the prisoners were disabled by mental illness and substance abuse. The lines for medications in the morning were extremely long. Only a relatively small minority of inmates—about 20 percent, I would say—appeared to be fundamentally criminal and callous, likely diagnosed with antisocial personality disorder, the hallmark of which is an uncaring attitude about the impact of one's actions on others. One could easily see the dynamics wherein the true criminals manipulated and schooled the more vulnerable and disabled prisoners. After two years at Osborne, I came away thinking that prisons were little more than taxpayer-funded schools of criminality. It was also clear that the system was essentially rigged against these men once they were released. It would be very hard for them to get a job: every employment application asks if you have ever been convicted of a felony. Many were released from prison without viable housing, and without follow-up medical and psychiatric treatment. In Connecticut, within three years of leaving prisons, almost 60 percent of ex-offenders were rearrested.

The research also said that the more vicious one's original crimes—the more violent and prolific and unfeeling—and the earlier they were committed, the far less likely an individual could be reformed. From what I had learned, there was little reason to believe that someone like William Outlaw could ever be rehabilitated. His criminal career began early, extended over a long time, and was highly violent. He came from multiple generations of criminals. He had been physically abused as a child. He had dropped out of high school. He had been incarcerated for a very long time. He had only received one of the treatments that had been shown to work—Richard Whitmire's group at Lewisburg was a form of cognitive behavioral therapy—but that was for just six months. His history was consistent with antisocial personality disorder. He was making just above minimum wage in his current jobs, and he had the skills and knowledge to make millions of dollars a year in other, illegal lines of work. In other words, there was nothing in the scientific research base that would give any indication that William Outlaw would—or even could—change.

But as I became more deeply immersed in the field of criminal justice rehabilitation, I found another line of research, research on "desistance," which is the factors that lead criminals, not to commit crimes (the focus of much of the prevailing research), but rather to reform their lives and exit from criminal careers. I read a book called *Making Good: How Ex-Convicts Reform and Rebuild Their Lives* by the prominent criminologist Shadd Maruna. Maruna asked individuals who had participated in long and entrenched criminal careers to tell him their life stories. He used an instrument called the "Life Story Interview," involving simple, almost literary questions: "Think of your life as a book or a movie; who are the main characters?" "What are some of the turning points?" "What's a high point?"

"What's a low point?" "What have you learned?" Around half of the people that Maruna interviewed continued on with lives of crime, while the other half had exited from what had once been prolific criminal careers. Maruna discovered that the two groups—which he called the "desisters" and the "persisters"—told their life stories in two different, and opposing, ways. The desisters recounted what Maruna called "redemption scripts," in which positive life events were emphasized over negative ones, and negative events were described in a way that stressed the positive lessons that might eventually arise out of them. Interestingly, Maruna found that the personalities of the desisters didn't change much, or at all: they were still the same people, but the stories they told about their lives had changed. The persisters, by contrast, told "condemnation scripts," which emphasized the blaming of others and a reluctance to take responsibility for their pasts. Desisters were not eager to blame themselves for their past mistakes either, but they were five times more likely to use active language in the way they told their stories. Desisters described a change process driven less by guilt and shame and more by the discovery—or rediscovery—of the good person that had been hidden inside of them, hidden by hard outer shells. In other words, reformed criminals could "re-biography" themselves, rewriting their pasts in order to live new futures. The ultimate and defining stage of recovery that Maruna found among desisters was "giving back," wanting to help younger people not make the mistakes they had.

I ended up contacting Maruna, now based in Northern Ireland. Over time we became friends and colleagues, coauthoring two articles and surviving a night of prodigious drinking with British criminologists at a conference at Oxford University. (I was surely a lightweight in this area.) In the meantime I founded a research institute at a large social service agency, The Connection, and became a desistance researcher myself. Midway through my work with Outlaw and partially inspired by him, I conducted, with colleagues, a Department of Justice–funded trial on the impact of "forensic peer mentors" in a halfway house program. We hired individuals with significant criminal histories who appeared to have turned their lives around—people like Outlaw—to counsel current offenders. Those residents who were assigned a mentor had a significantly lower (32 percent) criminal recidivism rate than those who did not have a mentor, who had a recidivism rate of 65 percent. This of course is a startling difference, and the project, and the program in which it was based, won a national award from a criminal justice association. Sadly, three subsequent applications to major foundations and the Department of Justice to continue the study were rejected.

It struck me that Outlaw was the classic desister. He had many of the attributes that Maruna uncovered: the self-efficacy, the belief in his own agency to control his future, a hugely optimistic mind, and a desire to give back to others. His behavior had changed but not his personality: he still wanted "money, fame and power" but this time he was acquiring it by doing good, or "making good," rather than bad. In many ways he was still the same person he always had been. He was still Juneboy.

I gave Outlaw Maruna's book. He devoured it. "That's me. The dude wrote about me. What can you say, Charlie, it just makes sense, what he wrote. Most people can't do it, but some people can: find an exit strategy from hell."

Based on his findings, Maruna and colleagues have proposed a new model for correctional practice, a "desistance model," which overturns decades of rehabilitation practice. Traditionally, correctional treatments are top-down and deficit-based—they apply treatments and interventions to "passive recipients who are characterized as deficient, ineffectual, misguided, untrustworthy, possibly dangerous, and almost certain to get into trouble again." By contrast, desistance-focused programs prioritize a person's strengths rather than risks or needs, creating opportunities for achievement, and are designed to enhance those things associated with lasting change, such as strong social bonds and involvement in prosocial activities. As colleagues of Maruna have written: "The distinction is crucial. Because desistance from crime may be associated with completely different factors from those that predispose a person to crime in the first place." Maruna expands on this: "Offenders might begin offending, in part at least, because of their impulsivity, failure to attend to consequences, preference for anti-social associates, unstructured lifestyles and emerging pro-criminal sentiments . . . and so on. But it doesn't follow that a reversal in these antisocial personality traits, behaviours and attitudes is what is key in moving offenders into desistance, or even in maintaining it." Maruna supports a recasting of those in offender management services "to think of themselves less as providers of correctional treatment (that belongs to the expert) and more as supporters of desistance processes (that belong to the desister). . . . We might be better off if we allowed offenders to guide us instead, listened to what they think might best fit their individual struggles out of crime, rather than continue to insist that our solutions are their salvation."

Here too Outlaw embodies the new paradigm. With the exception of Richard Whitmire and Mike Lajoie, nobody in the formal rehabilitation system guided him. What made him change was positive relationships, and new ways of seeing relationships, as a friend, a father, a son, and an employee; as a helper and healer rather than an exploiter. This allowed him to recast everything, including the way he saw violence. Outlaw puts it this way: "You have to tell your story. You have to give people hope. You have to be the CEO of you."

After I tracked down and left a message for Richard Whitmire, he happened to call back when Outlaw and I were meeting. Outlaw was overcome. He couldn't speak. All he could say was "Thank you, sir, for helping to save my life. Thank you, sir. Thank you. Thank you, Doc."

In February 2019, Outlaw did in effect become the CEO of himself. The contract for the New Haven Street Outreach program was put out to bid by the city. New Haven Family Alliance, which was undergoing management changes, chose not to bid. Leonard Jahad in the meantime had created his own nonprofit agency, CTVIP, standing for Connecticut Violence Intervention Program. Jahad and Outlaw wrote the grant and CTVIP was awarded the new contract. The City of New Haven, suffering from financial problems, reduced the size of the contract, but Jahad supplemented the budget by winning two new awards to do violence interruption, one for the City of Hamden, which adjoins New Haven, and one from Yale New Haven Hospital, in recognition for the work with families done by the team in the aftermath of shootings. Outlaw was issued a Yale New Haven Hospital identification card. He had, quite literally, gone from jail to Yale.

———

THE FIRST TIME that William and I ventured out into New Haven, in the spring of 2014, we went to Church Street South. Half-naked kids were throwing sticks at pigeons on the hill; two people pushed themselves around aimlessly in wheelchairs; and a number of the men and women who approached William were missing teeth and emaciated, clearly suffering from addiction. Two young men, former clients of Outlaw's at Goodwill, appeared and he gave them some suggestions for job opportunities. (One of them had just been fired from a job that Goodwill had found for him.) Only later did I realize that those conversations had occurred on the very ground on which Outlaw had killed Sterling Williams.

Four years later, a June 2018 headline in the *New Haven Register* read "Church Street South to be Razed": "A half-century after it was constructed, demolition permits were pulled Friday to raze the infamous Church Street South housing complex, where the last of its 266 families removed their belongings this week." In previous years, federal inspections had concluded the property—with its leaky roofs, crumbling porches, and pervasive mold—was unsafe. It scored twenty out of one hundred points on a HUD evaluation. After a multiyear tangle of lawsuits between the city and the owner, the city decided to knock Charles Moore's creation to the ground. A firm was hired at $5 million to oversee the process, which would last years. It takes two to three weeks to demolish a unit and dispose of the debris.

Just as there has been for a century, the city has grand plans for Church Street South. The latest vision is to build a thousand-unit mixed-income, mixed-use complex. However, the city was denied a federal application to build it for $30 million in 2016, and its second application in 2017 was turned down. But the city official in charge of the initiative, is optimistic: "As long as they keep putting [grants] out, we will go for it."

On a fall day, Outlaw and I went to the Jungle to observe the demolition. It would be the last time he would see his old territory. Over the intermittent sounds of bulldozers, we watched construction workers wrestle with piles of broken cinder blocks and contorted metal beams.

Most of the work seemed to involve people in hard hats standing around. William pointed at a surviving spray-painted doorway. "That was Sheila's apartment, where I first moved into the Jungle." He pointed to soon-to-be-doomed pine trees that were looking more sickly than ever. "That's where we had lines sometimes of thirty people waiting for the stuff. Soon when they're done, you will be able to see straight through to the train station and the police station. It will be wide open. People gonna be shocked at how big this place is."

Outlaw said he had a lot of unanswered questions as we stood there. "I can't deny I made a lot of money here, and had a lot of fun. I've been out for ten years and I'm healthy. But I had twenty-five years of bad times before that. A man died here. How can I reconcile those things? How do they even connect?"

I asked William if he thought the city was going to be successful in its plans for Church Street South. "Oh yes," he said with a surprising amount of certainty. "Yale wants it, Amtrak wants it. The police want it. Everybody wants it. They're gonna make something beautiful here."

Overlooking the rubble of the Jungle, February 2019.

ACKNOWLEDGMENTS

I would first like to thank William Outlaw for opening his life up to me. This has been an extraordinary experience for me, and I hope, for him.

I am forever grateful to Ivan Kuzyk and Pat Hynes of the Connecticut Department of Correction for alerting me to William's work in New Haven and introducing him to me.

I would like to thank the dozens of people I interviewed for this project, most of them multiple times: Leonard Jahad, Doug Bethea, Barbara Tinney, Pepe Vargas, Stacy Spell, Edward Kendall, Mike Lawlor, Michael Lajoie, Shadd Maruna, Kim Goodman, Mel Hall, Darryl McGraw, James Clark, Tom Ullman, Ira Grudberg, Patrick Clifford, Mary Loftus, Lisa Craggett, Gerald Scott, Bob Doucette, Andrew Papachristos, Brian Murphy, James Farnam, Mark Abraham and many others who shall remain nameless. A special thank you to Richard Whitmire for picking up the phone in Pennsylvania when I called him out of the blue one day. For critical and timely advice, I am deeply grateful to John Hall, Conrad Seifert and Beth Hogan.

I was blessed to have so many stalwart supporters who helped me through the length and intensity of this project. At Wesleyan's College of Letters and the Shapiro Writing Center, I thank Joseph Fitzpatrick, Sean McCann, Kari Weil, Anne Greene, Amy Bloom, and my friend and neighbor Andy Curran. I am indebted to my former writing teachers Phyllis Rose and the late Chris Kazan, and to my literary mentors Robert

Coles, and William Manchester and Paul Horgan at Wesleyan, who believed in me a long time ago. At The Connection, I was sustained on a daily basis by my brilliant and wonderful research colleagues, Michele Klimczak and David Sells. I will always be grateful to The Connection's CEO Lisa DeMatteis-Lepore and COO Beth Connor for their unremitting belief in me and their kindness. At Yale, I am in debt to my friends and colleagues Michael Rowe, Chyrell Bellamy and Mark Costa. Thank you to Jim Perakis for generously supporting my work.

I would like to acknowledge the extraordinary help and loyalty of my lifelong friends Mike Cianci of Cianci Creative LLC and Eve Marie Perugini, Jillian Frank, Pete Govert and June Plecan.

I have a dream team literary agent in Dan Conaway of Writers House. I can't believe that there's a better literary agent in all of New York City. Deep gratitude also to my film agent, Lucy Stille. At HarperCollins, I also have a dream editor in Denise Oswald, who always gave it to me straight (and calmly), particularly when she provided unerring control to a sprawling first draft. The whole team at Ecco—Dominique Lear, Miriam Parker, Martin Wilson, Meghan Deans—is superb. Thank you to Kim Tyler of Kim Tyler Photography, and Tyler Garzo of Sondroyo Communications who expertly created and manages my website.

For research and editorial support, and general good cheer, I thank Wesleyan students and former Wesleyan students, Nicole Updegrove, Natalia Siegel, and Ben Owen.

To me, coffee, music and writing are inseparable. I would like to acknowledge the myriad coffee places of central Connecticut—Dunkin Donuts, Blue State Coffee, Starbucks—where many of these pages were written, often while listening to artists whose rhythms inspire me: among them, Keith Richards, Joni Mitchell, Miles Davis, Steve Reich, Bartok, Pete Townshend, Robbie Robertson, and so many others

I would like to acknowledge my family above all. During the period in which I wrote this book, both my mother and father died, after having lived good and long lives. I will eternally be grateful that even in their declining health they heard draft material of this project. I knew I was on the right track when my father, a writer himself, smiled and gave me a thumb's up as I read the opening passages to him. "You've got it," he said. My mother always looked forward to hearing of my adventures with William. A lover of Proust and Stendahl, and also detective novels, she more than anyone else made me a writer. This book is co-dedicated to my brother John, as generous and smart a soul as I know. Tom, Marina, Marco, Alec, and Tyler: you sustain me. And to my wife, Laura, and son, Louis, are owed the greatest thanks: for putting up with my distracted moods for five years, for cheering me on, and for, both of them, having the biggest hearts in the world.

WILLIAM OUTLAW'S ACKNOWLEDGMENTS

I would like to acknowledge Charles Barber for coming into my life and making this project happen, and for the meaningful times we've spent together over the last five years. Above all, I would like to thank my children, my wife-to-be, and my mother, Pearl, for helping me through the hard times. I would also like to thank Tiger, Dapper Dan, Thomas Ullman, Leonard Jahad, Frank James, Warren Kimbro, Ivan Kuzyk, Patrick Hynes, Mike Lajoie, Mary Loftus, and "Doc" Richard Whitmire, all of whom believed in me when no one else did. Without their help, I would not be the person I am today.

FOR FURTHER INFORMATION

More information on William Outlaw and Leonard Jahad's Connecticut Violence Intervention Program, or CTVIP, can be found at www.ctintervention.org.

NOTES

PRELUDE: SEPTEMBER 24, 1988

xxi more than 40 percent of families of four live on less than $24,000 a year: "New Haven Divided by Growing Income Disparity," *New Haven Register*, February 22, 2014.

xxi New Haven was ranked as the fourth most violent city in the country: "The 10 Most Dangerous Cities in America," *The Atlantic*, May 26, 2011, https://www.theatlantic.com/national/archive/2011/05/the-10-most -dangerous-cities-in-america/239513/#slide7.

xxi every single victim of homicide: Paul Bass, "24 Murdered. None White," *New Haven Independent*, January 12, 2011.

CHAPTER 1: THE MONEY, THE FAME, AND THE POWER

8 the stabbing was justified: Tom Puleo, "Two Disturbances Bring Somers Prison 'Lockdown.'" *Hartford Courant*, October 29, 1991.

9 Oak City, North Carolina, population 500: "Population of Oak City, NC," https://population.us/nc/oak-city/, accessed May 8, 2019.

11 "imprisonment for life": Abbey Francis, "Long Lane School—The Early Years," *Wesleyan Argus*, September 16, 2013, http://wesleyanargus.com/2013/09/16 /long-lane-school-the-early-years.

CHAPTER 2: THE JUNGLE BOYS

20 20 percent of the population was forced to move: William Finnegan, *Cold New World: Growing up in a Harder Country* (New York: Random House, 1998), 34.

20 the red-line area of one city was characterized as having: "RACES: a Testbed for the Redlining Archives of California's Exclusionary Spaces": R. Marciano, D. Goldberg, C. Hou, http://salt.umd.edu/T-RACES.

20 In New Haven . . . the red-line districts included: "Mapping Inequality: Redlining in New Deal America," https://dsl.richmond.edu/panorama/redlining/#loc=13/41.3101/-72.9070&opacity=0.8&adview=full&city=new-haven-ct.

21 "a tangle of stress": Emily Dominski, "A Nowhere Between Two Somewheres: The Church Street South Project and Urban Renewal in New Haven," 2012, https://elischolar.library.yale.edu/mssa_collections/7.

41 5,000 to 7,000 intravenous drug users in New Haven: Andi Rierden, "Armed Youths Turn New Haven into a Battleground," *New York Times*, May 26, 1991.

43 33 percent of New Haven residents were employed in manufacturing: Amelia Nierenberg, "The Perimeter," *The New Journal at Yale*, October 16, 2017.

53 "People weren't worried about cocaine": Caleb Hellerman, "Cocaine: The Evolution of the Once 'Wonder' Drug," CNN, July 22, 2011, http://www.cnn.com/2011/HEALTH/07/22/social.history.cocaine/index.html.

54 cocaine market . . . valued at $140 billion: Llewellyn Hinkes-Jones, "How the Plummeting Price of Cocaine Fueled the Nationwide Drop in Violent Crime," *CityLab*, November 11, 2011.

54 "could afford to lose ninety percent . . . and still be profitable": Oriana Zill and Lowell Bergman, "Do the Math: Why the Illegal Drug Business is Thriving," https://www.pbs.org/wgbh/pages/frontline/shows/drugs/special/math.html, accessed October 14, 2018.

CHAPTER 3: KILL THE KING

73 "this case is innuendo and guessing": These quotes are taken from Attorney Ullman's notes at Outlaw's trial.

CHAPTER 4: RELEASE DATE: APRIL 7, 2073

89 "goals for our Latin community": Raymond Hernandez, "In Cities and Prisons, Hispanic Gang Grows," *New York Times*, November 29, 1992.

90 bodies of six people: Jesse Leavenworth, "A Grisly Night at New Britain's Donna Lee Bakery," *Hartford Courant*, April 2, 2014.

90 "They were friendly when they were sober": Mara Bovsun, "Death in the Doughnut Shop," *New York Daily News*, June 13, 2009.

99 "JB said to run him two bundles of weed": Edmund Mahony, "Crime Calls Collect As Inmates Phone Home," *Hartford Courant*, August 21, 1994.

100 "A team of 150": Andi Rierden, "On the Street: Drug Gangs Thrive as Arsenals Expand," *New York Times,* July 5, 1992.

101 "you'd have the National Guard in here": Edmund Mahony, "Agencies Apply More Legal Muscle Against Street-Gang Crime: Drug Arrests Show a New Approach," *Hartford Courant,* June 23, 1992.

105 the Interstate Compact: "Model Interstate Acts," *Journal of Criminal Law and Criminology,* vol. 26, no. 5 (Jan.–Feb. 1936): 773–85.

CHAPTER 5: BUS THERAPY

107 "stimulating qualities of the sunshine": Karen M. Morin, "'Security Here Is Not Safe'": Violence, Punishment, and Space in the Contemporary U.S. Penitentiary," *Environment and Planning D: Society and Space,* vol. 31, no. 3 (2013): 381–99.

111 "queers and steers": Most likely the officer was quoting a similar line from the film *An Officer and a Gentleman,* Paramount Pictures, 1982.

112 "fucked up everywhere else": Pete Earley, *The Hot House* (New York: Bantam Books, 1993).

113 brought under control by the National Guard: "Inmates Were Watching Murder Movie Before Riot," Associated Press, July 7, 1992.

115 most ferocious gang: David Holthouse, Intelligence Report, Southern Poverty Law Center, October 24, 2015.

133 "Mr. Persico, you're a tragedy": Arnold H. Lubasch, "Persico, His Son and 6 Others Get Long Terms as Colombo Gangsters," *New York Times,* November 18, 1996.

CHAPTER 6: HOW TO ESCAPE YOUR PRISON

141 "Anyone who is in power who is not willing to terminate, will be terminated": Robert W. Welkos, "Back in Touch with an 'Untouchable': New York Heroin Kingpin Leroy 'Nicky' Barnes, Who Operated in Plain Sight, Resurfaces in a Documentary," *Los Angeles Times,* October 21, 2007.

143 "30th birthday": D. Von Drehle, "Why Crime Went Away. The Murder Rate in America Is at an All-Time Low. Will the Recession Reverse That?" *Time,* February 22, 2010, 22–25.

145 "kill whatever D.C. Blacks and associates we could": United States v. Bingham, http://federalevidence.com/pdf/2011/08-Aug/US.v.Bingham.pdf.

145 exposing the ink to heat: Christopher Goffard, "Four in Aryan Brotherhood Guilty," *Los Angeles Times,* July 29, 2006.

145 "war with DC Blacks, T.D.": Christopher Goffard, "Aryan Brotherhood Mastered Low-Tech Network, U.S. Claims," Los Angeles Times, July 2, 2006.

145 "so he dies quick": "Defense Starts Today in Aryan Brotherhood Trial," *Orange County Register,* June 6, 2006.

158 money spent on federal corrections: "The Bureau of Prisons (BOP): Operations and Budget Report," Congressional Research Service, March 4, 2014, https://fas.org/sgp/crs/misc/R42486.pdf, accessed May 8, 2019.

CHAPTER 7: RETURN

160 40 percent of released offenders were rearrested in the first twelve months: "Recidivism of Prisoners Released in 30 States in 2005: Patterns from 2005 to 2010," Special Report, Bureau of Justice Statistics, Office of Justice Programs, U.S. Department of Justice, Washington, D.C.

160 thirteen times higher than that of the general public: Ingrid Binswanger, Marc Stern, et al., "Release from Prison—A High Risk of Death for Former Inmates," *New England Journal of Medicine* 356 (2007): 157–65.

CHAPTER 8: THE INTERRUPTER

180 "Please don't shoot": Colin Poitras, "Pleading 13-Year-Old Shot to Death: Unknown Shooter Sprayed Gunfire at Courtyard Neighbors," *Hartford Courant,* June 18, 2006.

180 "we will not tolerate it": Melissa Bailey, "Dixwell Mourns 'Nonnie'" *New Haven Independent,* June 22, 2006.

181 Justus Suggs . . . went into a coma: William Kaempffer, "Arrest Made in Teen's Slaying," *New Haven Register,* August 17, 2006.

181 a 133-page report: The Police Executive Research Forum, "New Haven Police Department Assessment Final Report," November 2007.

182 The institute's mission was based on the work of Martin Luther King: Nonviolence Institute, 2018, https://www.nonviolenceinstitute.org/mission-and-vision, accessed May 3, 2019.

185 Evaluations by the Centers for Disease Control and the National Institute of Justice: D.W. Webster et al., "Evaluation of Baltimore's Safe Streets Program: Effects on Attitudes, Participants' Experiences, and Gun Violence," Center for Disease Control and Johns Hopkins Bloomberg School of Public Health, January 2012; Wesley G. Skogan et al., "Evaluation of Cease-Fire Chicago," U.S. Justice Department, June 2009.

185 "his power to heal": C. G. Jung, *Collected Works of C. G. Jung,* vol. 16, *Practice of Psychotherapy* (Princeton: Princeton University Press, 2014).

190 "brought the level of violence down": Thomas MacMillan, "Newhallville-West River Truce Takes Hold," *New Haven Independent,* July 6, 2010.

191 "successful beyond expectations": Emily Bucholz, Georgina Lucas, and Marjorie Rosenthal, "Quantitative Evaluation of the New Haven Family

Alliance Street Outreach Worker Program," Robert Wood Johnson Clinical Scholars Program, Yale University School of Medicine, October 2009, https://www.ctdatahaven.org/sites/ctdatahaven/files/NewHaven%20 SOW%20Quantitative%20Evaluation%202009.pdf, accessed April 18, 2018.

191 intervened in 160 potentially violent disputes: Paul Bass, "Data Released on Project Longevity. Sort Of," *New Haven Independent,* December 19, 2014.

196 a third had seen someone shot or stabbed: "End Youth Violence," The Community Foundation for Greater New Haven, April 2015, https://www .cfgnh.org/About/NewsEvents/ViewArticle/tabid/96/ArticleId/101 /End-Youth-Violence.aspx, accessed May 8, 2019.

196 3 in 4 had heard gunshots: Alycia Santilli et al., "Bridging the Response to Mass Shootings and Urban Violence: Exposure to Violence in New Haven, Connecticut," *American Journal of Public Health* 107, no. 3 (March 1, 2017): 374–79.

198 Five thousand black men perish . . . annually: Violence Poverty Center, "Black Homicide Victimization in the United States: An Analysis of 2015 Homicide Data," http://vpc.org/studies/blackhomicide18.pdf, accessed April 28, 2019.

203 As he put it, "I get shot": Andrew Papachristos, personal correspondence, May 2, 2019.

203 He discovered 4,107 such cascades in Chicago: Ben Green, Thibaut Horel, and Andrew V. Papachristos, "Modeling Contagion Through Social Networks to Explain and Predict Gunshot Violence in Chicago, 2006 to 2014," *JAMA Internal Medicine* (March 1, 2017): 326–33.

CHAPTER 9: THE MAINTENANCE OF HOPE

215 The numbers in 2017: Paul Bass, "Homicide Rate Hits 50-Year Low," *New Haven Independent,* January 2, 2018.

216 eighth most murderous city: "The 30 Cities with the Highest Murder Rates in the US," *Rapid City Journal,* November 13, 2017.

222 "I heard good things about New Haven": Paul Bass, "Police Chief Brings Reform Message to Obama," *New Haven Independent,* October 23, 2015.

222 "crime in Connecticut is at a 48-year low": "Connecticut's Second-Chance Society," editorial, *New York Times,* January 4, 2016.

223 reducing short-term violence: Michael Sierra-Arevalo, Yanick Charette, and Andrew V. Papachristos, "Evaluating the Effect of Project Longevity on Group-Involved Shootings and Homicides in New Haven, CT," Institution for Social and Policy Studies, October 2015, https://isps.yale

.edu/sites/default/files/publication/2015/10/sierra-arevalo_charette
_papachristos_projectlongevityassessment_isps15-024_1.pdf.

POSTSCRIPT: ABOUT THIS PROJECT

242 an instrument: Dan McAdams, "The Life Story Interview," https://www
.sesp.northwestern.edu/foley/instruments/interview/, accessed April 18,
2019.

244 "desistance model": S. Farrall, *Rethinking What Works with Offenders*
(Cullompton, UK: Willan Publishing, 2002); S. Farrall, "Social Capital and
Offender Reintegration: Making Probation Desistance Focused," in *After
Crime and Punishment: Pathways to Offender Reintegration,* ed. S. Maruna
and R. Immarigeon (Cullompton, UK: Willan Publishing, 2004).

244 By contrast, desistance-focused: Shadd Maruna and Thomas P. LeBel,
"The Desistance Paradigm in Correctional Practice: From Programs to
Lives," in *Offender Supervision: New Directions in Theory, Research and
Practice,* ed. Fergus McNeill et al. (Cullompton, UK: Willan Publishing,
2010), 65–89.

244 "The distinction is crucial": M. K. Harris, "In Search of Common Ground:
The Importance of Theoretical Orientations in Criminology and Criminal
Justice," *Criminology and Public Policy* 4 (2004): 311–28.

244 "But it doesn't follow": F. Porporino, "Bringing Sense and Sensitivity to
Corrections: From Programmes to Fix Offenders to Services to Support
Desistance," in *What Else Works? Creative Work with Offenders,* ed. J. Bray-
ford, F. Cowe, and J. Deering (eds.) (Cullompton, UK: Willan Publishing,
2010).

246 "As long as they keep putting [grants] out, we will go for it": Markeshia
Ricks, "The Tear-Down Begins," *New Haven Independent,* June 25, 2018.

INDEX